Contents

The Scholarship Series in Biology

General Editor: W. H. Dowdeswell

The Mechanism of Evolution

THE SCHOLARSHIP SERIES IN BIOLOGY

Human Blood Groups and Inheritance
S. Lawler and L. J. Lawler

Woodland Ecology
E. G. Neal

New Concepts in Flowering Plant Taxonomy
J. Heslop Harrison

The Mechanism of Evolution
W. H. Dowdeswell

Heathland Ecology
C. P. Friedlander

The Organization of the Central Nervous System
C. V. Brewer

Plant Metabolism
G. A. Strafford

Comparative Physiology of Vertebrate Respiration
G. M. Hughes

Animal Body Fluids and their Regulation
A. P. M. Lockwood

Life on the Sea-shore
A. J. Southward

Chalkland Ecology
John Sankey

Ecology of Parasites
N. A. Croll

Physiological Aspects of Water and Plant Life
W. M. M. Baron

Biological Principles in Fermentation
J. G. Carr

Studies in Invertebrate Behaviour
S. M. Evans

Ecology of Refuse Tips
Arnold Darlington

Plant Growth
Michael Black and Jack Edelman

Animal Taxonomy
T. H. Savory

Ecology of Fresh Water
Alison Leadley Brown

Ecology of Estuaries
Donald S. McLusky

Variation and Adaptation in Plant Species
David A. Jones and Dennis A. Wilkins

Nature Reserves and Wildlife
Eric Duffey

Enzymes and Equilibria
J. C. Marsden and C. F. Stoneman

The Mechanism of Evolution

Fourth Edition

W. H. Dowdeswell, M.A., F.I.Biol.

Professor of Education, Bath University

Heinemann Educational Books

Heinemann Educational Books Ltd

LONDON EDINBURGH MELBOURNE AUCKLAND TORONTO
HONG KONG SINGAPORE KUALA LUMPUR
IBADAN NAIROBI JOHANNESBURG
LUSAKA NEW DELHI

ISBN 0 435 61251 4

First published 1955
Second edition 1958
Third edition 1963
Reprinted 1964, 1967
Fourth edition 1975

Published by
Heinemann Educational Books Ltd
48 Charles Street, London W1X 8AH

Printed in Great Britain by
The Whitefriars Press Ltd
London and Tonbridge

Preface to Fourth Edition

In writing this book, I have assumed that the reader will have an elementary knowledge of genetics and general biology such as is obtained in a school sixth form or first year university course. It has been gratifying to find that the previous edition has proved useful for undergraduates, and with their needs particularly in mind I have expanded somewhat the scope of the references in the Bibliography, including journals that are likely to be available in a university library.

During the last decade or so, great advances have been made in the study of evolution; indeed, it is no exaggeration to say that a whole new discipline has come into being–that of ecological genetics. To take such changes into account has meant a radical recasting of the pattern and contents of this book, and in producing the present edition the opportunity has also been taken of adopting the new format of the Scholarship Series.

While assembling the material for this new edition, I have become increasingly aware of the debt I owe to an ever widening circle of friends, not least to the many students and teachers with whom I have worked, both in school and university. My particular thanks are due once again to Professor E. B. Ford, F.R.S., with whom I have had the privilege of collaborating for many years and who has been a continual source of valuable advice and encouragement. It is, perhaps, worth placing on record that the foundations of our joint studies of ecological genetics were laid on the remote Scottish island of Cara as long ago as 1938. Since then, work with the meadow brown butterfly, *Maniola jurtina,* has contributed appreciably towards an understanding of the mechanism of evolution and many of our findings are summarized in this book. I also owe much to Dr E. R. Creed and Professor K. G. McWhirter who have contributed so greatly to the work on *Maniola* for many years.

It is also a pleasure to express my thanks to those who have kindly supplied illustrations for this book. Mr S. Beaufoy provided the photographs for Figures 20, 29, 31, 36, 44, and the cover picture; Mr Eric Hosking Figure 33; Dr H. B. D. Kettlewell Figure 48; Dr A. J. Salsbury Figure 27; Mr Colin Wilson Figure 16, Dr M. Ashby provided the original roughs on which are based Figure 30; Mr W. E. S. Powers Figures 4, 7, 17, 24, 25, 43, 50, 53, 54, and 55; and Mr R. Ward Figure 3. My sincere thanks are also due to my secretary, Mrs E. Thomas, for her unfailing patience and efficiency in the preparation of the typescript and the index.

1975 W. H. Dowdeswell

1

Introduction

It is now generally accepted that organisms living today have arisen from a primitive ancestral stock through a process of continual change extending over great periods of time. This concept of organic evolution is a fundamental part of modern biology, for without it, any idea of taxonomy or the comparative study of animals and plants would be almost meaningless. Of recent years a welcome trend in biology has been towards a more interdisciplinary approach in which evolution has provided an important unifying theme. Evidence of the new thinking is provided by the increased numbers of courses both in schools and universities covering biology as a single subject rather than treating botany and zoology as separate areas. Within the subject itself, disciplines that have traditionally remained somewhat isolated from each other are now exhibiting an increased awareness of the fundamental linkages that exist between them. For instance ecology, genetics, evolution, and mathematics have proved to be a particularly fruitful alliance when brought together within the general compass of ecological genetics, and it is with this area of biology that much of this book is concerned.

One of the problems of studying evolution is that much of the evidence frequently put forward in its support appears to be circumstantial. The only tangible grounds for assuming that such organic changes have occurred in the past derive from fossils and, as Darwin so clearly pointed out, the fossil record is far from complete. Another problem facing students of evolution is the huge time factor involved; it has been suggested that, on average, a million years may be needed for the formation of a new species.

However, if we look below the species level, we can detect a dynamic process at work producing changes in plant and animal populations at a speed that is susceptible of scientific investigation. Indeed, there is now every reason to believe that the rate of change in living organisms today is at least as fast as during the 750 million years since life first appeared on the earth's surface. Viewed as a whole, we can do no more than evaluate the evidence for the main evolutionary trends throughout these great vistas of time—sometimes referred to as macro-evolution.

But if we are to consider in more detail the *mechanism* by which evolutionary changes are brought about, we must inevitably look for these events in their ultimate location. This is among populations in their habitats—woods, fields, hedgerows, rivers, or even, for some parasites, within a single animal or plant. Modern scientific methods, including mathematics, have provided us with effective tools for gauging minute evolutionary changes—micro-evolution, and it is with such changes that we will be concerned in the chapters that follow. To study them is of great importance, for in doing so we are at once confronted with an active process in full operation at the present moment. Viewed thus, in conjunction with evidence from the past, the whole evolutionary picture assumes a more realistic and convincing form.

Before embarking on a discussion of modern theory and experimentation, it is essential to have a clear idea of the foundations on which the subject is based and of our indebtedness to the pioneers in this field. During the last 200 years scientific thought has undergone vast changes, and nowhere have these been more apparent than in the sphere of evolutionary biology. No account of our present ideas would be complete without some reference to the great contributions of Darwin, the men who preceded him, and other famous biologists of the mid-nineteenth century. The value of their work and its significance today will be the subject of the next chapter.

1

Historical Aspects of Evolution

Man has speculated on the possibility of organic evolution ever since the dawn of historic time and possibly long before it. In ancient Persian and Egyptian mythology we find recurrent ideas of anthropomorphic gods arising from primordial matter who proceeded to create the universe. At the point in early Indian thought where mythology merges into philosophy, Brahma was conceived as the eternal Being from whom all living things evolved materially as well as spiritually, the process being one of transformation rather than of original creation. The first ideas on evolution which bear any resemblance to our own are to be found among the ancient Greeks in the teachings of a group of Ionian philosophers known as the Milesians. Typical among them was Anaximander (610-547 B.C.) who believed in the existence of indeterminate primordial matter as the basis of cosmic and organic evolution. That this material was watery, he concluded from fossil evidence, since he observed that the sea had originally extended over a much larger area than it covers today. But it is not until Heraclitus (*c.* 510 B.C.) that we find in embryonic form the introduction of another modern concept, the idea of *conflict* among living organisms and a struggle between them for survival. As we shall see later, in outline this bears some resemblance to Darwin's theory of Natural Selection. It was against this background of thought that Aristotle (384-322 B.C.) made his famous contributions which were to remain almost unchallenged for 2000 years after his death.

In attempting to assess the contributions of the early Greeks, we must remember that their views were firmly grounded in the idea of spontaneous generation and the fixity of species. It is therefore not surprising that Aristotle's principal contribution in the field of evolution was his system of classification—an attempt to bring order into the *chaos* of living organisms that he observed so acutely. The details of Aristotle's taxonomic system need not concern us here. Suffice it to say that the classificatory criteria he adopted were mostly quite arbitrary, such as the possession by animals of red blood, their number of legs, and the presence or absence of wings. In modern terminology, his system was *artificial* rather than *natural.* By dividing living organisms into even smaller groups according to their more minute resemblances, Aristotle was able to construct his *scala naturae* ('ladder of nature'), a graded series with man at the top, and primitive plants followed by inanimate matter at the bottom.

While Aristotle was particularly concerned with animals, his pupil Theophrastus (*c.* 380-287 B.C.) made similar contributions in the field of botany. For instance, the taxonomic system he propounded distinguished clearly between Monocotyledons and Dicotyledons on the basis of the number of embryonic leaves and the arrangement of their veins.

The legacy of the early Greeks by way of classification was destined to survive with very little change until the seventeenth century, when advances in our knowledge of plant and animal anatomy made possible a more critical approach and heralded in the age of the great systematists. Among them, one of the earliest and most distinguished was the Englishman John Ray (1628-1705). Combining the latest discoveries in plant and animal anatomy with the traditional classificatory system of Aristotle, he worked out many of the larger plant and animal groups as we know them today. The basis of Ray's system was mainly detailed anatomical structure combined with evidence from geographical distribution (see page 12). As such, it represented the first attempt at a *natural* classification. As an ordained minister, Ray was a fundamentalist in outlook and accepted as true the biblical account of creation. Nonetheless, he was well aware of intraspecific variation in plant species which he attributed entirely to environmental causes, such as the fertility of the soil, variations in temperature and the like. Ray's views on the fixity of species are well summarized in the following passage, 'Plants which differ as species preserve their species for all time, the members of each species having all descended from seed on the same original plant' (Stearn, 1957).

The death of John Ray almost coincided with the birth of the Swede Karl Linnaeus (1707-1778) who was destined to become one of the predominant influences in eighteenth century biology. Much of his work consisted in the elaboration and development of the ideas of Ray and his contemporaries, particularly as regards the smaller groupings of organisms–classes and orders. But the contribution for which Linnaeus will always be remembered was at the lowest level of the classificatory system–the concept of the genus and species, leading to the system of *binominal nomenclature* used universally today. Much of the Linnean classification still remains unchanged, indeed the tenth edition (1758) of his great work *Systema Naturae* has been adopted by universal agreement as a source of reference for the international standardization of species names.

It is often stated that, as with John Ray, the possibility of evolutionary change never occurred to Linnaeus and his followers, who conformed to the religious doctrine of the day in accepting the idea of special creation and therefore the fixity of species. However, from some of his later writings it is clear that Linnaeus was well aware of the existence of variation within species which he explained as being due to hybridization. Writing in 1762, he points out that since all Genera are primeval, there must, in the beginning, have been as many Genera as there were individuals. Later, these organisms became fertilized by others of different kinds giving rise to the diversity of species and varieties that we know today.

The first signs of a fundamental change in outlook are to be found among the eighteenth century French biologists. Thus Buffon (1707-1788), noted for his remarkable *Histoire Naturelle* in forty-four volumes, pointed out that factors such as predation and competition for survival might cause one species to replace another in the course of time and to undergo a change of appearance in the process. Buffon propounded no theory to explain how such changes might be brought about or perpetuated. Nevertheless, he can be justly credited with the introduction of a new mode of thought moving away from the static outlook of Linnaeus towards the more dynamic approach characteristic of the beginning of the nineteenth century.

The ideas of Buffon evidently impressed Erasmus Darwin (1751-1802), grandfather of Charles Darwin, for he discusses in his *Zoonomia* the possibility that climatic and other environmental variations might cause far-reaching changes in the appearance of plants and animals subjected to them. For instance, the white coats of many arctic mammals were supposed to have arisen in response to the influence of their snow-clad surroundings, a view which came close to anticipating the doctrine propounded by Lamarck.

Contributions of Lamarck (1744-1829)

Lamarck's ideas on evolution are to be found in the greatest of his works–*Philosophie Zoologique* published in 1809. They have been summarized concisely by Wilkie in the form of four 'laws':

(i) Nature tends to increase the size of living individuals to a predetermined limit;
(ii) the production of a new organ results from a new need;
(iii) the development reached by organs is directly proportional to the extent to which they are used;
(iv) everything acquired by an individual is transmitted to its offspring.

In attempting to apply his 'laws' Lamarck found himself in a somewhat paradoxical situation. In accordance with his belief in a Divine plan of creation involving perfection at every level of elaboration, he subscribed to the idea of a continuously graded series of living things ranging from the simple to the highly complex. The existence of obvious gaps in the series he attributed to a lack of human understanding which would be remedied in time. On the other hand Lamarck was fully aware that evidence both from living organisms and fossils strongly suggested that the various groups of plants and animals had radiated out tree-wise during the course of the earth's history, giving rise to new forms adjusted to a wide diversity of environments. This process we now know as *adaptive radiation* (see also page 131).

One prime requirement for the operation of such a process was a mechanism by which the advantageous characteristics possessed by an individual could be, as it were, stamped on its hereditary constitution and so passed on to its descendants. This led Lamarck to postulate the theory of 'the inheritance of the effects of use and disuse'. Carried to extremes we have the example of the ancestral snake which took to a crawling habit and so lost its limbs, and the ancestral giraffe whose neck became progressively elongated by generations of browsing on trees. In essence Lamarck's theory was merely a special aspect of the idea of the inheritance of acquired characters propounded by Erasmus Darwin. Both believed that changes in organisms resulting from the influence of their surroundings could be inherited, but whereas Lamarck was concerned solely with the effects of use and disuse, Erasmus Darwin regarded the environment in general as being responsible for inherited change, as did Buffon before him.

Judged in a modern context, both these theories were open to the same objection; they assumed a mechanism of evolution which operated by *controlling mutation.* Now, no means is known whereby the effects of the natural environment can be stamped on the chromosomes carried in the nuclei of the germ cells, nor does it seem theoretically likely that such a thing could occur. But even if it did, the amount of variation produced would be many times too small to account for the observed rate of evolution.

Lamarck's theory was, however, a very difficult one either to prove or disprove. In the first place, its exponents could always take refuge in the immense periods of time necessary for detectable changes in an organism to become apparent. Secondly, any experimental approach demands a pure line of animals or plants from which inherent genetic variation can be eliminated. This is difficult to obtain in practice, and in all probability has never been used in any of the experiments so far performed. An alternative would be to study a population subjected only to environmental effects with all possibility of artificial or natural selection eliminated. Again, this would be a problem of great complexity. On grounds of common sense we might expect that if some external influence were to cause a modification in a large proportion of a population at each generation, visible results should be obtained in the course of the kind of experiments which could be conducted under artificial conditions. On the other hand, as Haldane has pointed out, if only a few individuals are affected it can be shown mathematically that such slowly acting causes could never succeed in establishing in wild populations characteristics of more than neutral survival value.

The question whether characteristics acquired during the life-time of an organism can be inherited, still has a certain contemporary relevance in view of the claims made by T. D. Lysenko's school of geneticists in Russia. Their assertions are difficult to evaluate in Western terms due to the ideological overtones associated with them. Judged superficially, some undoubtedly suggest the possibility of the inheritance of acquired modifications, but caution must be exercised in interpreting such results, as has been well demonstrated by Waddington. By administering a temperature shock of 40 °C to the pupae of a stock of the fruit fly, *Drosophila melanogaster,* he has shown that a disturbance in the development of the wing-veins occurs, giving rise to a condition known as *cross-veinless.* This effect is apparent in some individuals but not all. Cross-veinless flies were selected for breeding, and after their pupae were given the heat shock the proportion of abnormal flies was found to rise at each generation (Figure 1). From the fourteenth generation onwards, a few cross-veinless flies occurred among the untreated members of the selection line in the absence of the environmental stimulus required by the original population. Such a result is susceptible of a Lamarckian

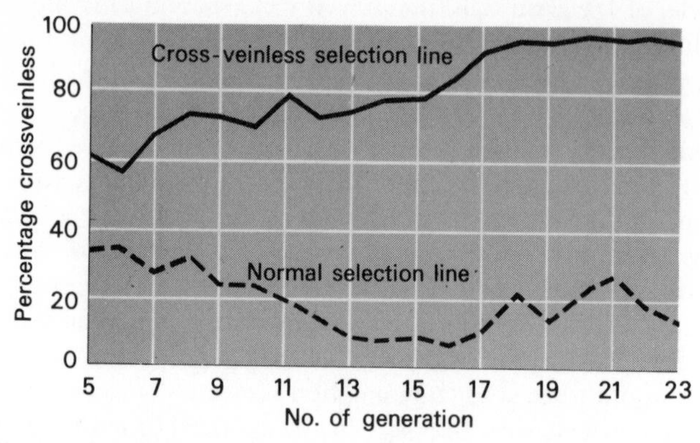

Figure 1. Experimental selection for and against the ability of *Drosophila melanogaster* to produce a cross-veinless phenotype in response to temperature shock. (After Waddington.)

interpretation, namely that a modification initially induced by the environment later became incorporated in the hereditary constitution of the organism. However, returning to a point made earlier, entirely homogeneous genetic material is virtually impossible to obtain, and in this instance the stock used was certainly heterogeneous. Evidently, none of the initial population carried a genotype able to produce a cross-veinless phenotype at the normal temperature (25 °C) but some were able to do so at 40 °C. Once this had occurred, selective breeding could have served to enhance the effect. The fact that some genotypes gave cross-veinless individuals at 25 °C can be explained simply in terms of the gene complement that they happened to possess. Thus the original population can be judged to have contained two genotypes:

(i) giving normal venation, with and without heat shock;
(ii) giving normal venation with no heat shock and cross-veinless with it.

To these must be added, as a result of selection, a further genotype,

(iii) giving cross-veinless with and without heat shock.

Malthus and the theory of populations

Before we come to consider the work of Charles Darwin, we must note the contributions of several men who in their different ways helped to condition contemporary Victorian thought for the acceptance of his ideas. One of these was Dr T. P. Malthus (1766-1834) a distinguished mathematician and economist who is remembered chiefly for his *Essay on Population,* published in 1798. In this he made various disturbing predictions concerning the trend of world population in relation to food supply, pointing out that while the former might be expected to multiply in geometrical progression, increases in the latter, at best, could never be more than linear (Figure 2). Hence, he considered that the limit

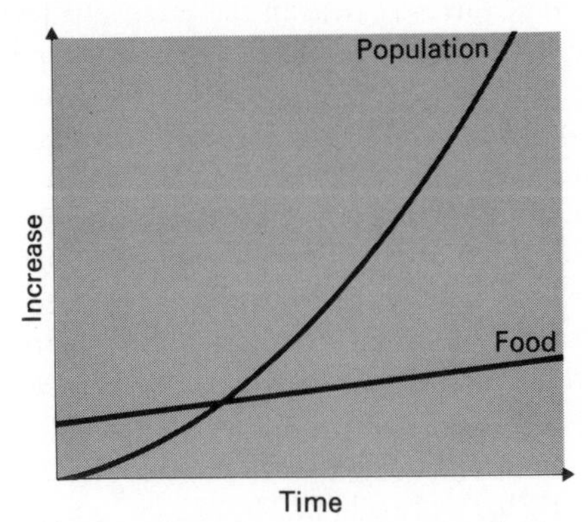

Figure 2. The relationship between population and food supply as seen by Malthus

of man's numbers must ultimately be governed by the availability of his food, and unless some artificial check were imposed upon his rate of multiplication he would inevitably breed himself towards starvation. A worldwide struggle for survival would then ensue in which the less well endowed races and individuals would be bound to suffer and ultimately perish. As we now know, Malthus's predictions, although reasonably accurate as far as the rate of man's increase was concerned, were not fulfilled because of the contributions to world resources which resulted from the opening up of the New World. Nonetheless, his ideas, centering as they did round the concept of competition and the control of numbers, undoubtedly helped to pave the way for the theory of natural selection.

Contributions of contemporary geologists

A characteristic feature of the history of biology during the early part of the nineteenth century was its close association with that of geology. In a sense, both were passing through similar phases. The great French biologist Cuvier (1769-1832) had appreciated the significance of fossils and realized that they represented a kind of succession involving changes in some forms of life and the disappearance of others. When studying the rocks in the vicinity of Paris, he was struck by the fact that some strata exhibited fossils of a particular kind, others adjacent to them differed completely in their fossil content, while some contained none at all. From this discontinuity, Cuvier concluded that a series of catastrophes must have occurred at various stages in the earth's history, the latest being the flood recorded in the Book of Genesis. After each catastrophe living organisms had repopulated the earth, partly through further creation and partly from those living elsewhere that had managed to avoid disaster. Thus, underlying Cuvier's theory was the idea of extinction, which represented a new way of thinking in both biology and geology. However, contemporary English geologists such as James Hutton (1726-97) took a more realistic view, and noted that the changes occurring among fossils occupying successive strata were always gradual and showed no evidence of the erratic fluctuations that Cuvier's theory demanded.

But the rise of modern geology can be said to stem largely from the contributions of Sir Charles Lyell (1797-1875) whose famous three-volume work *Principles of Geology* (1830-33) was destined to exert a profound effect on contemporary thought, not only in geology but in biology as well. Darwin records that on his exploratory voyage in H.M.S. Beagle he took the first volume and that it 'was of the highest service'. Lyell's appreciation of the significance of fossil succession served to impart a new precision to the dating of strata in sedimentary rocks. Most important of all from the biological viewpoint, he showed that the various forces which had been responsible for the laying down of successive rock formations were still acting in precisely the same manner as in the past. There was, therefore, no longer any need to postulate catastrophes. All that was required to explain geological succession was a continuous and universal process of erosion and sedimentation. Such views at once made sense of the fossil record and came to be known as the doctrine of *uniformity*. Paradoxically, Lyell used his theory at first to argue against catastrophism and therefore against the idea of evolution in living things. Darwin, on the other hand, saw in uniformity a principle which could account as much for biological innovation as for geological change. Unlike Cuvier, who imagined that changes took place in a series of leaps, Darwin viewed them as part of a slow and continuing process.

While Lyell is rightly regarded as the first uniformitarian in geology, Darwin undoubtedly occupied the corresponding niche in biology.

Darwin's Theory

The groundwork of our present knowledge of evolution was laid by Charles Darwin (1809-82), an outstanding naturalist with acute powers of observation and deduction in the field. In 1831 he embarked upon a voyage in the naval frigate H.M.S. Beagle which was engaged in making geographical surveys in South America. Darwin travelled in the capacity of unpaid naturalist and the wealth of observations that he made were destined to prove the turning point in his career. By the time of his return five years later he was convinced of the reality of evolution and had accumulated a mass of evidence in support of his views. This evidence can be summarized briefly as follows:

(i) *Recent geological changes in South America.* It will be remembered that Darwin was deeply influenced by Lyell's views and he soon realized that the uniformity postulated by the geologists was equally applicable to palaeontology and hence to the study of evolutionary succession. Numerous visits to points along the South American coast provided striking evidence in support of this view. On 20 February 1835, Darwin recorded his sensations during a short earthquake. A few days later, on making measurements of the strata in Conception Bay, he found that the land had risen by two or three feet. Not far away, putrefying mussels were discovered still clinging to rocks now some 3 m above high-water mark.

(ii) *Fossil evidence.* While in South America, Darwin unearthed the remains of many fossil animals–monkeys, rodents, and ruminants, some being of vast size, such as the elephant-like *Toxodon,* equipped with a dentition for gnawing. Many of these bore a striking resemblance to species still alive; for instance the fossil remains of the giant mammals (Glyptodonts) were similar in general appearance to modern armadillos (Figure 3) although they differed from them in detail, particularly size. Darwin concluded that although distinct from one another, the two groups must be closely related and that divergence must have occurred in comparatively recent geological time.

(iii) *Geographical succession.* During the various calls along the East Coast of South America Darwin collected and observed many kinds of

Figure 3. (a) Modern armadillo (length about 1 m) and (b) a reconstruction of a Glyptodont (length about 4 m)

local animals and plants. He noticed that individuals of a particular species varied somewhat from one locality to the next but the closer they were together, the less was divergence apparent.

(iv) *Geographical variation.* While in the Pacific, a visit was made to the Galapagos, a group of volcanic islands lying on the Equator (Figure 4), about 600 miles west of Ecuador and 1000 miles south-west from Panama. This was destined to influence Darwin profoundly, for here he observed that the flora and fauna of the islands were different from, yet clearly related to those of the South American continent. Moreover, a similar situation prevailed among the various islands in the group. He found that the inhabitants could judge from what island an individual of a species of giant tortoise came by observing small

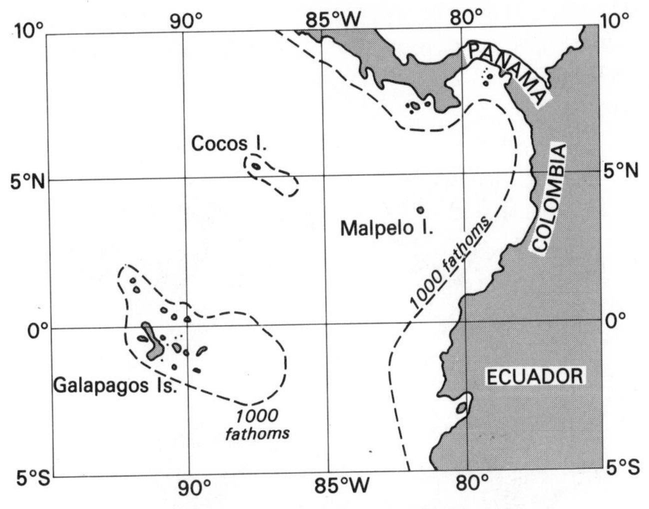

Figure 4. The Galapagos Islands

peculiarities in its shape, for instance whether the carapace was more or less circular or oblong. Darwin's work was mainly concerned with the local birds and he found that the finches, in particular, exhibited characteristic variations. Thus, beak size and shape were beautifully adapted to the respective diets of the different species (Figure 5) ranging from the formidable bills of *Geospiza*—a seed-eating genus, to the predominantly insect-eating *Camarhynchus* and *Certhidea.* Apart from such first-order variation, smaller second-order differences were also discernible when samples of the colonists of different islands were compared. Thus Lack (1938) obtained measurements of such features as depth of beak and breadth of wing in samples of the same species on different islands. Some of his data are summarized in Table 1.

From such observations Darwin concluded that nearly related species must have been descended from a common interinsular stock. Were two such species to meet in the same region they would presumably have tended to compete, and both would have survived only by becoming isolated from each other by their food requirements or some other ecological characteristics. The more divergent the existing forms, the longer and more effective must their isolation have been. Presumably, the founder individuals must have colonized the islands direct from the South American continent. The fact that the finches showed such a remarkable range of diversity no doubt suggested to Darwin that their ancestors must have been among the earliest arrivals to colonize the Galapagos Islands.

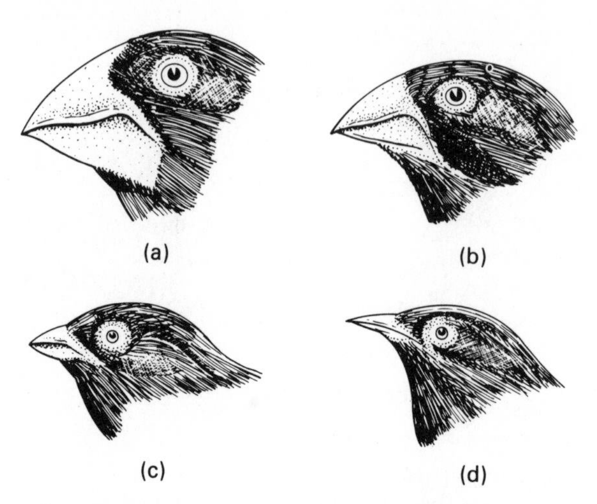

Figure 5. Heads of four species of Galapagos ground-finches showing variation in beak structure correlated with diet: (a) *Geospiza magnirostris;* food large seeds, (b) *G. fortis;* food smaller seeds. (c) *Camarhynchus parvulus;* insectivore, and (d) *Certhidea olivacea;* insectivore (small insects)

Table 1. Mean measurements (mm) of beak depth and wing breadth in males of the ground finch, *Geospiza,* on different islands of the Galapagos. The average sample size was about 45. (After Lack.)

Island	*G. magnirostris*		*G. fortis*		*G. fuliginosa*	
	Beak	Wing	Beak	Wing	Beak	Wing
Wenham	20.4	86	–	–	–	–
Tower	21.2	86	–	–	–	–
Abingdon	20.0	82	11.8	67	7.7	60
Bindloe	19.1	82	12.1	67	7.4	59
James	20.5	84	12.5	72	8.0	64
Jervis	18.6	83	12.6	–	8.2	64
Indefatigable	19.1	–	12.8	73	8.2	64
Duncan	–	–	11.5	70	8.1	64
North Albemarle	–	–	12.4	71	8.1	64
South Albemarle	18.4	–	13.9	75	8.3	65

Soon after his return to England, Darwin read Malthus's essays and immediately extended his ideas to cover all organisms living under natural conditions. He argued that those best adapted to the environment in which they live must, on the whole, survive and breed at the expense of the less well adapted, which will tend to be eliminated in the struggle for survival. This was the theory of *natural selection.* It is one of the most remarkable coincidences in the history of science that at the same time another English naturalist, A. R. Wallace (1823-1913), then exploring the Malay Archipelago, should have come independently to precisely the same conclusion as Darwin regarding the mechanism of evolution. The problem of publishing their findings simultaneously was overcome with ingenuity and tact. Following the joint intervention of Lyell and the botanist J. D. Hooker, both papers were communicated together to the Linnean Society on 30 June 1858. Shortly afterwards, in 1859, Darwin published his famous work *The Origin of Species by Means of Natural Selection, or The Preservation of Favoured Races in the Struggle for Life* (generally known as *The Origin of Species*). The work of Lamarck and, later, that of Lyell and his contemporaries, had set the stage for the acceptance of Darwin's views. The existing climate of opinion on evolution, coupled with a widespread interest in natural history, largely account for the remarkable enthusiasm with which the book was received by the educated public of the Victorian age.

But besides an abundance of organisms and the forces of natural selection, Darwin realized that one other requirement was necessary for evolution to take place. His extensive studies of animals and plants both in domestication and under natural conditions had led him to appreciate the unending variety of living things, and it soon became clear to him that in the variability of species lay their potentiality for evolution. In other words, *variation* was the raw material on which natural selection must act in order to bring about the adaptation of organisms to their diverse and constantly changing environments. Moreover, the mere existence of variation was not enough; there must also be an effective means of transmitting characteristics both good and bad from one generation to the next. From his earliest days Darwin had been impressed with the 'force of inheritance' and he assembled many observations on characteristics in domestic animals and plants which had appeared spontaneously in breeding and had been perpetuated in their offspring. He was later to publish this information in 1868 under the title, *The Variation of Animals and Plants under Domestication.* At that time there existed no comprehensive theory of inheritance, for although

Mendel was then working at Brünn in Moravia and had published his famous findings in 1866, thirty-four years were to elapse before their significance was appreciated and publicized by Correns, De Vries, and Tschermak in 1900.

Darwin realized that some structural mechanism must exist which is present throughout the bodies of living organisms and responsible for the transmission of inherited characteristics. This led to his hypothesis of Pangenesis in which he postulated the existence of hereditary 'gemmules' transmitted from one generation to the next in the gametes at sexual reproduction. The details of Pangenesis need not concern us here, but the idea exerted an important influence on the contemporary theory of evolution. Darwin believed that, as a general rule, the characteristics of parents merged together in their descendants, thus losing their original identity. He adds, however, that 'some characters refuse to blend, and are transmitted in an unmodified state either from both parents or from one'. This process of *blending inheritance* can be expressed mathematically as follows.

For any species subject to variation there must be a specific mean for an inherited character or set of characteristics, that is to say, their average occurrence judged from random samples of the particular species. If x and y represent the respective deviations from this mean of two parents, disregarding the effects of mutation, the deviation of the offspring will be equal to $\frac{1}{2}(x + y)$. In practice, the variability of organisms is now generally expressed as their *variance,* being the mean value of the square of the deviation, in our example x^2 or y^2. Hence, in blending inheritance, the variance of the offspring will be the mean value of $[\frac{1}{2}(x + y)]^2$ or $\frac{1}{4}(x^2 + 2xy + y^2)$. This implies that with each successive generation the fund of inherited variability will be reduced, the trend being towards ever increasing uniformity–the reverse situation of that required for the effective operation of natural selection.

Darwin's field observations had led him to believe that all living organisms are subject to spontaneous inherited changes, and he regarded these changes as giving rise to two distinct kinds of variation. Large and easily recognizable variants he called 'sports', which were essentially of a clear-cut (discontinuous) kind. 'Fluctuating variation', however, was less obvious or extreme, and nearer the continuous variation that we recognize today, such as human height. It was this that Darwin regarded as the raw material on which natural selection must act if evolution were to take place. Herein lay the fund of latent variation which served as a reservoir for that continually lost through blending inheritance. As Sir

Ronald Fisher pointed out, the fact that variation would be rapidly reduced by blending at each generation means that the bulk of it present at any moment must be of extremely recent origin—less than one-thousandth can be ten generations old. Hence, if any beneficial variety derived from mutation was to be used in artificial breeding or by natural selection, it would have to be selected instantly before becoming diluted by blending. Although apparently quite clear on this point, Darwin was constantly troubled by the paradox, knowing evolution to have been a slow process involving slight and gradual changes.

Another problem which puzzled Darwin was the great variability observed among domesticated species both in their native homes and in the countries to which they had been transported. This he explained as being at least partly due to the improved conditions of feeding—a distinctly Lamarckian approach. It seems certain that Darwin realized the necessity for postulating some other means of inheritance by which the hereditary material passed on at each generation could retain its identity and accumulate. But no concrete scheme ever emerged in his time, and it was not until 1900, when Mendel's work was rediscovered, that the missing link in the chain of evolutionary theory became apparent.

Modern theory of evolution

Since Darwin's time the study of the mechanism of evolution has followed three main courses:

(i) A critical examination of Lamarckism in the light of modern knowledge of physiology and genetics. While it cannot be asserted that the theory has been fully disproved, for reasons already given, there is no convincing evidence of its operation in any instance cited so far. The last of the Lamarckist revivals took place in Russia in 1948 under the leadership of Lysenko, but the claims of his school still remain to be substantiated.

(ii) The determination of the true sources of variation and of the means by which both advantageous and disadvantageous characteristics can be transmitted uncontaminated from generation to generation. This has been made possible by the great advances in genetics during the last fifty years.

(iii) A more objective approach to the study of natural selection and, in particular, an appreciation of the value of variable species in experimental work.

The first of these has already been considered in some detail and there is no need to carry the discussion further. Concerning the second, the advent of Mendelism provided just the impetus necessary to establish Darwin's theory and to gain its general acceptance at a time when widespread doubts had begun to arise regarding its universal application. We have seen how blending inheritance proved a stumbling block at the time of *The Origin of Species. Particulate Inheritance,* on the other hand, provided an ideal mechanism for perpetuating variance. In the first place, Mendel had shown that the genetic factors (*genes*) do not contaminate one another but, with certain rare exceptions (*mutation*), retain their identity in successive generations. Furthermore, it was now clear that, as a result of the random assortment and recombination of genes, subject to the restrictions imposed by linkage and crossing over, an almost unlimited fund of variation was available in all organisms reproducing sexually. Thus if two genes controlling the same character in a wild population breeding at random exist in the proportion of p to q, the three possible genotypes will appear in the ratio $p^2 : 2pq : q^2$ (see page 29). In fact, these proportions are seldom achieved under natural conditions, partly on account of chance survival of certain individuals rather than others and, more particularly, because of the effects of natural selection. For it is likely that different genotypes interacting with the same physical environment will give rise to phenotypes adapted to varying extents to the circumstances in which they occur. Hence those which are best suited will stand a better chance of survival and reproduction than the rest.

This concept leads us to another important discovery of Mendel's, the fact of *dominance.* R. A. Fisher showed conclusively that an inevitable outcome of the sort of selective process outlined above will be that those genes which are beneficial will tend to become dominant to their less advantageous alleles. Moreover, the speed with which this process takes place will depend on the magnitude of the selective advantages involved; hence it plays an important part in influencing the results of natural selection.

Another advantage of particulate inheritance was that it removed the necessity for postulating a high mutation rate. Darwin had been forced to adopt this idea, probably against his better judgment, because of the effects of blending in reducing variation. We now know that the incidence of gene mutations is far lower than had previously been supposed and, as we shall see in the next chapter, it is desirable for evolutionary progress that this should be so. It has been estimated that the fund of variance in sexually reproducing organisms due to past mutations is now so enormous that, even if no further mutations were to occur, there is every reason to believe that as a result of Mendelian inheritance evolution could continue indefinitely at the same rate as in the past.

The detailed study of natural selection is of much more recent origin, and it is only during the last forty years or so that precise methods of analysis have been employed to estimate the nature of selective agencies and the circumstances in which they act. For this we are largely indebted to Fisher, a brilliant exponent of Darwinism who devised many of the specialized mathematical techniques necessary for research in this field.

To summarize the modern outlook, we may say that the essential features of Darwin's theory still hold good. But with the advent of Mendel and subsequent research into the nature of particulate inheritance, our views on the origin and maintenance of variations have greatly changed. The existence of natural selection is still accepted today but, as a result of much experimental work both in the field and in the laboratory, our knowledge and outlook are more sophisticated than when *The Origin of Species* was published. Early attempts at studying its speed and magnitude were largely unsuccessful on account of badly designed experiments and inadequate mathematical techniques. More precise methods have been developed only in comparatively recent years. With Darwin, we accept the fact that the *variation* of living organisms and *natural selection* acting upon them together constitute the basis of evolution. An appreciation of the relative roles of these two processes is vital for the understanding of present-day theory, and they will be discussed further in subsequent chapters.

2

Variation: its Origin and Continuity

Contribution of Mendelian Inheritance

We now know that the great bulk of inheritance in both animals and plants is of a Mendelian kind and that any other hereditary mechanisms which may exist play only a subsidiary part. Recent years have seen vast strides in our knowledge of 'molecular biology', particularly in relation to the microstructure of chromosomes and the mechanism of gene-action. Chromosomes are composed principally of the nucleic acid DNA (desoxyribose nucleic acid) whose molecule consists of two long chains of sugar-phosphate units (a single chain may be 10 000 units long) wound round one another spirally, forming a pair of helices. The cross-linkages between the two chains are formed from the bases adenine, thymine, cytosine and guanine, which are so arranged that adenine on one chain is always linked to thymine on the other, similarly cytosine with guanine. Hence, it follows that although the sequence of bases on any one chain may vary, their order automatically determines that of the opposite chain which must be complementary (Figure 6). We can now see why the DNA molecule is said to be 'in code', for the four bases, although obeying a fixed 'law' governing their combination with one another, are subject to immense variation in the order in which they occur. The complement of chromosomes found in the nucleus of every living organism must thus carry a unique sequence of A, T, C, and G units in their DNA molecules, and hence be capable of transmitting a great diversity of genetic information.

The question remains, how is this information communicated from the nucleus in a form which can be interpreted by the surrounding cytoplasm? While virtually the whole of the DNA contained in a cell is concentrated in the nucleus, the related compound RNA (ribose nucleic acid) is found partly in the nucleus but principally in the cytoplasm. Now it has long been known that the various metabolic processes associated with cytoplasm are dependent upon the catalytic action of

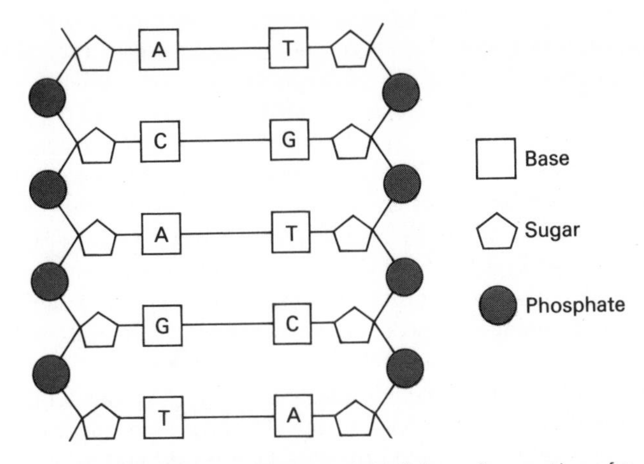

Figure 6. Diagrammatic representation of the linkages in a portion of a DNA molecule. Note that the sequence of bases in one helix is complementary to that of the other

enzymes. It is therefore significant to find that the cell's rate of synthesis of these substances is closely correlated with its RNA content. Thus, while the cells of heart muscle and kidney synthesize relatively little enzyme and contain small amounts of RNA, those of pancreas and liver are rich in RNA and produce great quantities of enzyme secretions. It therefore appears that RNA acts as the messenger which transmits the genetic code from nucleus to cytoplasm, no doubt passing through the pores in the nuclear membrane which electron microscope studies have shown to exist. The sequence of events outlined above can be summarized roughly as follows:

nuclear DNA → nuclear (messenger) RNA

→ cytoplasmic (transfer) RNA → enzymes ⇄ metabolic processes

The foregoing summary is no more than a bare outline of our present knowledge of chromosomes and how they act. Nor is it possible here to consider modern theories concerning the nature of the gene although brief reference is made to this in Chapter 7.

The principle that genes act through enzymes which influence the biochemical processes going on in the cytoplasm around them is well illustrated in the flour moth, *Ephestia kühniella.* In the typical wild form the eyes of the adult and skin of the larva are brown owing to the formation of pigment. A single mutant recessive (*v*) gives rise in the homozygous condition to a pink-eyed adult and an almost colourless larva. If organs from a wild larva are transplanted into a colourless one during the penultimate instar, normal dark insects are produced. A similar result is obtained by injecting an alcoholic extract of the tissues of a normal larva. It has been shown that the formation of pigment in *Ephestia* is dependent upon the presence of the enzyme kynurenine, an intermediate amino-acid product of tryptophane metabolism. Furthermore, the experimental treatment of 'albino' larvae has confirmed expectation, namely, that the depth of colour is directly proportional to the dosage of enzyme. Thus the physiological action of the gene *v* is to inhibit the synthesis of kynurenine and hence to deprive the insect of the catalyst necessary for the production of pigment.

As we shall see shortly, there are many genes which exert more profound physiological effects, such as influencing growth rate, and these have undoubtedly played an important part in the evolution of a number of species, including man.

It has long been known that not all genes are carried in the nucleus; some are cytoplasmic. The inheritance of leaf colour in *Primula sinensis* is entirely maternal and therefore non-nuclear, as is shown by the following crosses:

♀		♂		Offspring
yellow	×	green	→	yellow
green	×	yellow	→	green

The genes concerned are carried in the plastids and hence are known as *plastogenes.* Sometimes the situation is more complex, as in some species of the evening primrose, *Oenothera,* where two green-leaved individuals may, in certain circumstances, give yellow-leaved offspring. Thus:

♀		♂		Offspring
O. muricata (*curvans*) (green)	×	*O. hookeri* (green)	→	green
O. hookeri (green)	×	*O. muricata* (*curvans*) (green)	→	yellow

The resulting nuclei must be identical in both crosses, but the cytoplasm of the fertilized eggs will be different, since more will have been contributed by *O. muricata* in the first cross and by *O. hookeri* in the second. Hence, interaction between the hybrid nuclei and factors carried in the cytoplasm must decide the eventual colour of the offspring. This is an instance of the combined action of nucleus and cytoplasm. Cytoplasmic genes are also known in animals (e.g. *Paramecium*), where they appear to be freely distributed in the cells and not associated with any particular body or inclusion. They are therefore known as *plasmagenes*.

It has been suggested that factors carried in the cytoplasm may be largely responsible for determining the broad fundamental characteristics of phyla, classes, and orders, but at present there is little evidence to support this idea. It is certain, however, that such non-Mendelian inheritance plays little part in producing the inherent variation necessary for the effective operation of natural selection under wild conditions.

Mendel believed that each gene exerted one specific effect, and that heredity thus depended on unit factors controlling unit characters. We now know that this is not so, but that while the genes retain their identity they produce their effects as a result of joint action. This was proved as long ago as 1905, when Bateson and Punnett studied the inheritance of rose and pea combs in fowls (Figure 7), both of which are

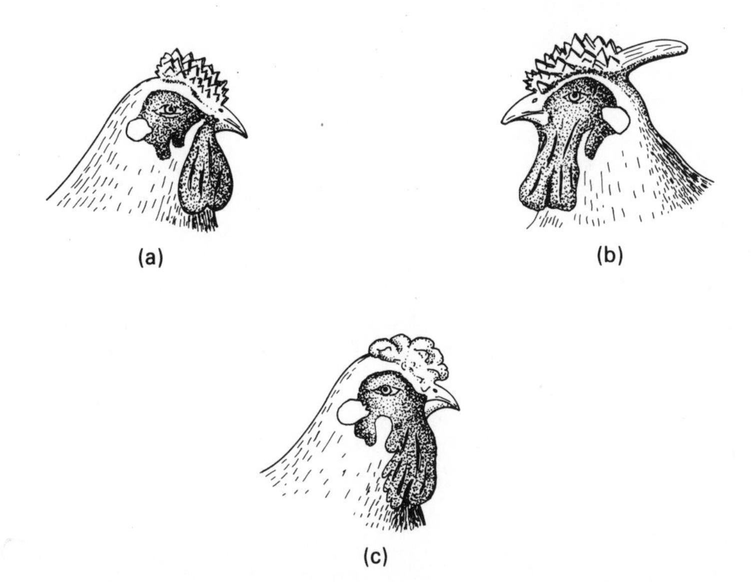

Figure 7. Comb shape in poultry: (a) pea, (b) rose, and (c) walnut

dependent upon unit factors dominant to the normal single comb. They found that crossing rose- and pea-combed fowls produced a new type, 'walnut', as a result of the interaction of the two genes.

Many examples of this kind are now known from which we see that the genetic equipment of an organism must be regarded, from the point of view of its effects, as a single entity or *gene complex*—the internal environment formed by the genes themselves. Its composition will be subject to constant change due to the recombination of factors in particulate inheritance and, to a relatively minute extent, the process of mutation. The phenotypic result is variation, and since this will be inherited it is known as *genetic variation.*

Climatic changes of quite small magnitude have been shown to exert surprising effects on the expression of genes under experimental conditions. No doubt this is also true in the wild state. Among plants, for instance, an albino strain in barley has been shown to be due to the action of a single recessive. This only exerts its full effect below a temperature of 6.5 °C, while above 18 °C no chlorophyll reduction occurs at all and the genotyptic albinos are indistinguishable from the normal form.

Similar situations occur in man. For example, the potentiality for freckling is under genetic control but the extent to which freckles actually develop depends upon the action of the ultraviolet component of sunshine. Thus, in those carrying the appropriate genes, an outdoor life will lead to more freckling than one passed mainly in artificial light.

Changes with no genetic association which result from the action of external factors, such as variations in diet or muscular exercise, are known collectively as *environmental variations;* these are not inherited.

Sometimes it is possible to demonstrate the interaction of genotype and environment in influencing the expression of a gene. In the fruit-fly, *Drosophila melanogaster,* the gene 'antenna-less' results in the absence of one or both of the fly's antennae according to its degree of expression. Among laboratory stocks some normal flies exhibit no ill effects even when carrying the gene in the homozygous state. In flies emerging early[1] the effect is less in the males than in the females; the reverse is true later on. In both sexes the effects are at a minimum about the fourth day of emergence and reach a maximum at eight days. It has been shown that

[1] An insect is said to emerge when it appears from the chrysalis (pupa) as a complete adult. The period of emergence refers to the duration of this process among the individuals of a species living under particular conditions.

the environmental factor concerned is the food of the larva. Changes due to the deterioration of the artificial culture medium result in a corresponding increase in the expression of the antenna-less gene. Vitamin B_2 is apparently a vital factor, for when it is present in adequate amounts all flies exhibit the ill effects to a reduced extent, and some are unaffected.

Patterns of variation

The work of Mendel, like many of his successors up to the present day who have been involved in the breeding of plants and animals, centred round a study of clearly contrasted characteristics. Some of those used by Mendel himself in his experiments with the garden pea (*Pisum*) are summarized in Table 2.

Table 2. Some characteristics of the garden pea (*Pisum*) studied by Mendel

Structure	Dominant	Recessive
Stem	Long	Short
Leaves (cotyledons)	Yellow	Green
Flowers	Axillary	Terminal
Fruits (pods)	Inflated	Constricted
	Green	Yellow
Seeds	Round	Wrinkled
	Yellow (endosperm)	Green (endosperm)

Using the last two seed characters, the mechanism of inheritance is summarized in Figure 8. Mendel's results are shown in Table 3.

Table 3. Results of some of Mendel's crosses of the garden pea (*Pisum*)

	Yellow Round	Yellow Wrinkled	Green Round	Green Wrinkled
Seeds obtained	315	101	108	32
Experiment ratio	9.8	3.2	3.4	1
Theoretical ratio	9	3	3	1

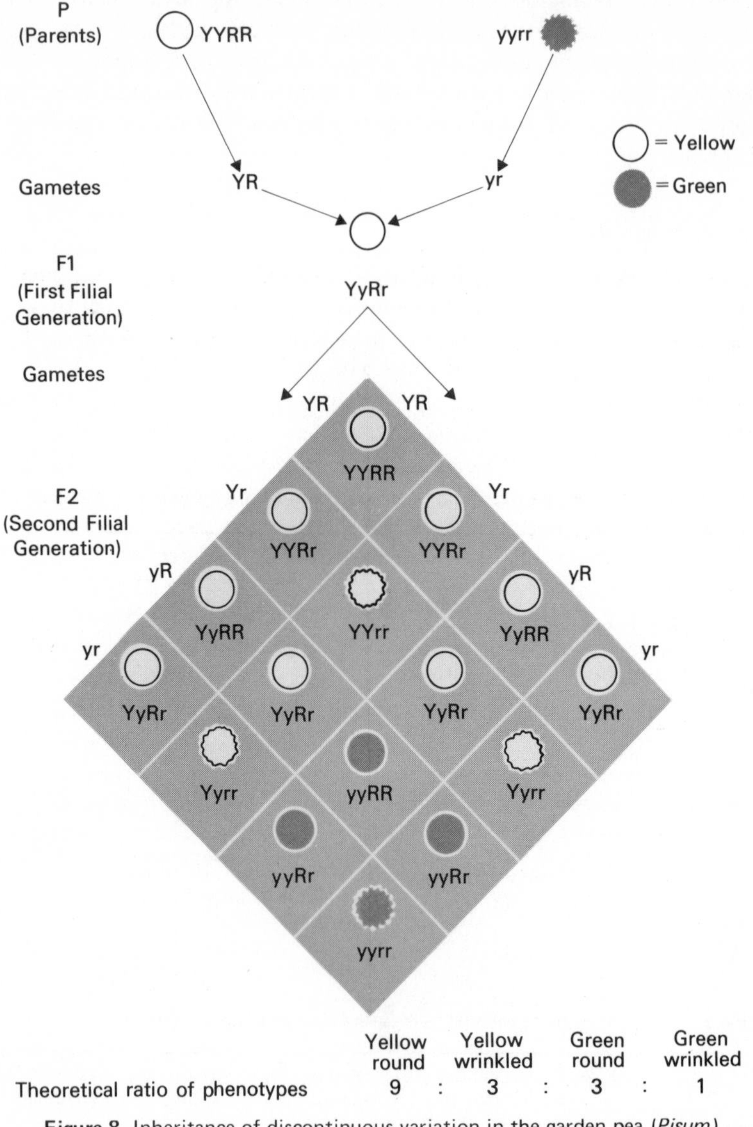

	Yellow round		Yellow wrinkled		Green round		Green wrinkled
Theoretical ratio of phenotypes	9	:	3	:	3	:	1

Figure 8. Inheritance of discontinuous variation in the garden pea (*Pisum*)

Since the pairing of gametes is at random, the probability that the experimental ratio will attain the theoretical value increases with the number of zygotes formed. In this instance the number of seeds produced (556) was quite large and the agreement with expectation fairly close. Variation of this kind involving clear-cut and contrasted differences is known as *discontinuous*.

One of the reasons for the slow acceptance of Mendel's findings was that many influential biologists such as Pearson and Galton were pursuing a biometrical approach to heredity, and studying variants such as human height which exhibit not discontinuity but a graded range of variation. To them the concept of blending inheritance provided a ready explanation of *continuous variation* which Mendelism apparently did not.

We now know that this view was mistaken and in fact many alleles, for instance those controlling colour variations in animals, are additive in their effects, so that their degree of expression depends on the number present in the genotype. Let us take as a theoretical example a species of moth whose wing colour varies from black through grey to white. Pigmentation is controlled by two pairs of unlinked alleles, the dominants B_1 and B_2 producing pigment, the recessives b_1 and b_2 inhibiting it. The depth of coloration therefore depends on the number of doses of B (maximum 4). Figure 9 summarizes the results of crossing two double heterozygous individuals, the colour-range and frequency of the resulting phenotypes being recorded in the form of a histogram. Since more than one pair of alleles is involved in controlling the same character, this kind of inheritance is known as polygenic. Like discontinuous variation, the continuous form is explicable in strictly Mendelian terms, and in order to account for a wide range of diversity we need only invoke the action of relatively few pairs of genes.

Variation in populations

So far we have considered variation only in relation to the results of gene segregation in specific crosses. But under natural conditions populations are composed of many interbreeding families, each contributing its quota of genes to a common *gene pool*. The extent to which a particular allele is present or absent in an interbreeding group may well exert a profound

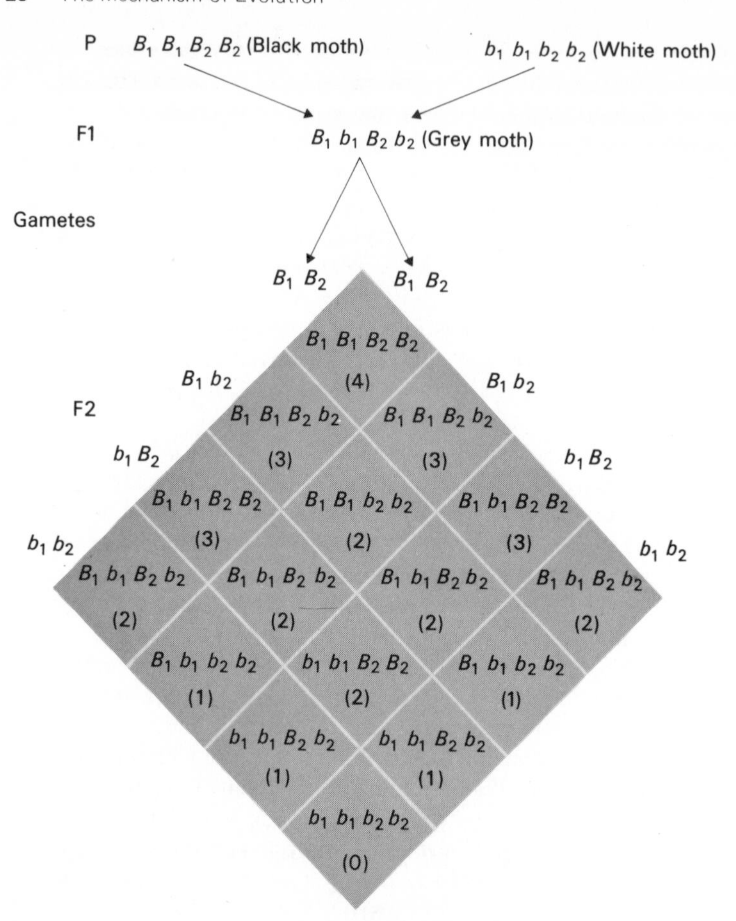

Figures in brackets denote the number of pigment alleles in each zygote

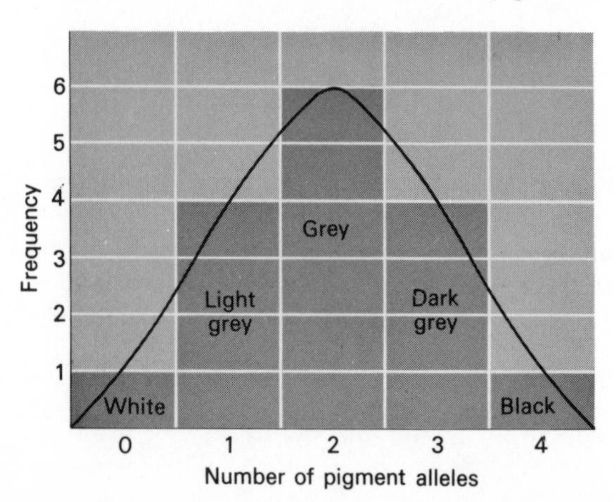

Figure 9. Inheritance of continuous variation (theoretical example)

influence on its degree of adaptedness in a given situation. Hence, when studying population genetics, we are particularly concerned with determining *gene frequencies.* In 1908 an English mathematician, G. H. Hardy, and a German physician, W. Weinberg, showed that, while the proportions of dominant and recessive genes in populations may vary, in the absence of any forces causing a change in gene frequency the relative proportions of each allele will remain constant in successive generations.

To take a hypothetical example, let us suppose that we are studying a population with a gene pool containing the alleles A and a. The three possible genotypes are, therefore, AA, Aa, and aa.

Let the frequency in the gene pool of $A = p$ and the frequency in the gene pool of $a = q$. Then $p + q = 1$ (i.e. 100 per cent of the population).

Knowing the frequency of the two alleles, we can determine that of the three genotypes they will produce:

(i) *genotype AA.* This can only result from the union of two gametes each carrying A. Hence the probability of such a union (i.e. its frequency) will be $p \times p = p^2$.

(ii) *genotype aa.* The situation will be similar to (i), the genotype frequency being $q \times q = q^2$.

(iii) *genotype Aa.* Here, the A-bearing gametes may be provided by either the male or the female, and similarly for the a-bearing gametes. There are therefore two possibilities,

$$(A \times a) = p \times q = pq$$

$$= 2pq$$

or

$$(a \times A) = p \times q = pq$$

Since the three genotype frequencies must add up to 1 (i.e. 100 per cent), we can write,

$$p^2 + 2pq + q^2 = 1$$

This is known as the Hardy-Weinberg equation.

Its method of use can best be illustrated by a simple example. In the snail *Cepaea hortensis,* a pair of alleles (U and u) control the banding of the shell, that causing a lack of bands being dominant. In a certain population 9 per cent of the shells are banded. What proportions of the population are heterozygous and homozygous for U? The relevant information can be summarized as follows:

Genotypes	*UU*	*Uu*	*uu*
Phenotypes	Unbanded	Unbanded	Banded
Proportions	p^2	$2pq$	q^2
Percentages	49	42	9

The figures for percentages are derived as follows:

The genotype frequency of $uu = q^2 = 0.09$ (9 per cent). Now by Hardy-Weinberg, $p + q = 1$, and so

$$q = \sqrt{0.09} = 0.3 \text{ (= gene frequency of } u)$$
$$p = 1 - 0.3 = 0.7 \text{ (= gene frequency of } U)$$

Hence
$$p^2 = (0.7)^2 = 0.49 \text{ (49\%)}$$
$$2pq = 2 \times 0.7 \times 0.3 = 0.42 \text{ (42\%)}$$

Thus, in order to maintain 9 per cent unbanded shells, 42 per cent of the population must be heterozygous for *U* and *u*.

As was mentioned earlier, the Hardy-Weinberg equilibrium can only obtain provided certain conditions are fulfilled:

(i) the population must be large enough to exclude errors due to non-random sampling,
(ii) random mating must take place between all genotypes,
(iii) individuals of the three genotypes must have equal chances of survival (i.e. there must be no selection) and reproduction (i.e. they must have the same fertility),
(iv) mutation rates must be low,
(v) there must be no immigration or emigration (i.e. the population must be reproductivity isolated).

Although the equation applies to a rather idealized situation it can, nonetheless, be of great value as a model in providing an approximation. Moreover, on occasions where it is possible to detect the effects of an allele in both the heterozygous and homozygous states (see Chapter 4), it provides a means of assessing the degree of divergence from an expected ratio and hence of assessing the effect and magnitude of selection.

Mutation

Mutations are defined as any heritable variations other than those resulting from segregation and recombination of genes. They are ultimately responsible for all evolutionary novelties and may arise in four distinct ways.

(i) *Chromosome multiplication.* The number of chromosomes in the cell nucleus of a particular species of animal or plant is often regarded as constant. As far as is known, this is true for most animals and the majority of plants, although in both the chromosome number sometimes fluctuates considerably. These fluctuations are due to irregularities in cell division resulting from a number of causes which are not fully understood. Their occurrence can be stimulated artificially by treatment with such substances as colchicine, a procedure now widely used in horticulture for the production of new plant varieties, e.g. in tomatoes. Instead of the normal somatic number of chromosomes (*diploid* or $2n$), the resulting organism may have three times the haploid number (*triploid*), four times (*tetraploid*), or even higher numbers still. Such individuals are known as *polyploids.* Their fertility is often much reduced compared with diploids because of difficulties in chromosome pairing at meiosis, since a single chromosome may find itself with more than one homologue. Polyploids are particularly well known among plants and account for many important cultivated varieties (cultivars), for instance in roses and cereals.

Again, variations in chromosome number may be due to the loss or gain of one or more chromosomes giving rise to *aneuploids* in which the different numbers are not simple multiples of one another. The most usual way in which this situation is brought about is through non-disjunction, when one or more pairs of homologous chromosomes fail to separate during meiosis and pass to the same pole. In the resulting daughter cells one will lack a chromosome and the other will have an extra one (Figure 10).

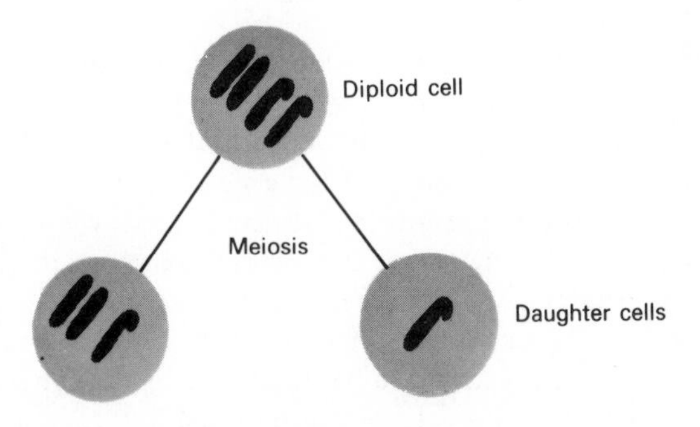

Figure 10. Non-disjunction. Failure of one pair of chromosomes to separate at meiosis results in daughter cells with different chromosome complements

(ii) *Chromosome fragmentation.* The number of chromosomes may remain constant while alterations in structure take place in one or more of them. Thus a portion of a chromosome may become detached and lost (*deletion,* Figure 11), or it may be broken off, reversed, and

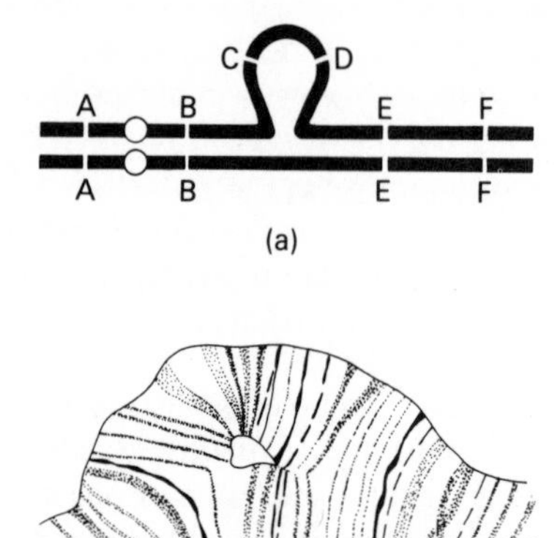

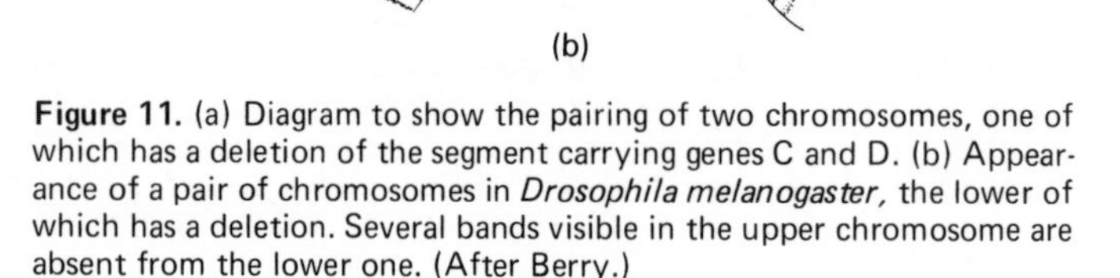

Figure 11. (a) Diagram to show the pairing of two chromosomes, one of which has a deletion of the segment carrying genes C and D. (b) Appearance of a pair of chromosomes in *Drosophila melanogaster,* the lower of which has a deletion. Several bands visible in the upper chromosome are absent from the lower one. (After Berry.)

subsequently joined on again (*inversion*). Again, it may become attached to the homologous chromosome (*duplication*) or to a non-homologous chromosome (*translocation*). In each instance the result will be a rearrangement in the linear order of the genes, and hence a source of genetic variation. It is important to distinguish such events from the normal process of crossing over which takes place between *chromatids* derived from different but homologous chromosomes and leads to the formation of the chiasmata characteristic of the diplotene stage of the first meiotic prophase. Chromosome fragmentation can be induced experimentally by such agencies as mustard gas, X-rays, and other ionizing radiations.

Dobzhansky has made an extensive study of the effects of certain chromosome inversions in various species of the fruit fly, *Drosophila,* making use of the large banded chromosomes occurring in the salivary glands. The fact that the frequency of these inversions is characteristic for each population and varies from one locality to another, also at different times of the year, indicates that they must confer balanced advantages and disadvantages which are subject to the control of natural selection. This view has been substantiated by laboratory experiments where one form has been shown to be superior to another in relation to such factors as temperature tolerance and the availability of food. The occurrence of the various inversions is only detectable by cytological means and it is not known at present what phenotypic effects they exert. Unlike the situations considered so far, the control mechanism is not a single pair of alleles, but a group of closely linked genes operating together and acting as a kind of switch-mechanism. Such a gene-system is known as a *super-gene*.

(iii) *Repetition.* Study of salivary-gland chromosomes in *Drosophila melanogaster* has revealed that certain sections possessing a characteristic kind of banded appearance may occur in two different parts of the same chromosome (Figure 12). These reduplicated portions of chromatin

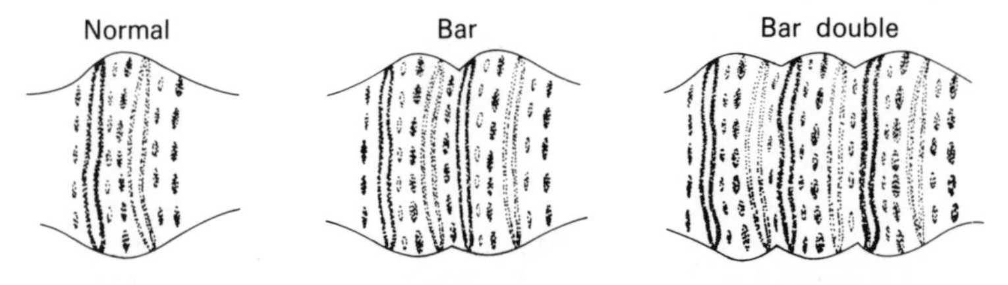

Figure 12. Repetition in *Drosophila* at the Bar eye locus. (After Sutton.)

(sometimes known as *repeats*) are found to exert some mutual attraction and tend to pair in the diploid condition. This further suggests a fundamental similarity between them and the possibility that they carry allelic genes. As Dobzhansky pointed out, while a normal diploid insect carries most of its genes in duplicate, those borne on the repeat sections of the chromosomes will be represented four times. Repetition has now been detected in a number of species of *Drosophila,* but its study has not extended much beyond the giant salivary gland chromosomes of the Diptera.

Although the phenotypic effects of the reduplicated genes so far detected could have little survival value under wild conditions, the process is nonetheless of great potential importance in evolution. It differs from chromosome fragmentation, already discussed, in one important respect, namely that, apart from polyploidy, it is the only mechanism known whereby an increase in the number of genes possessed by an organism can be achieved.

(iv) *Gene mutation.* Occasionally a spontaneous change occurs at a gene locus without any corresponding effect on the number of chromosomes or their structure. A new allele is thus formed. In some species a number of these gene mutations have been detected at the same point and the result is known as a *multiple allele series.* Only two members of the series can of course be represented in the cells of any one diploid individual (see Figure 13).

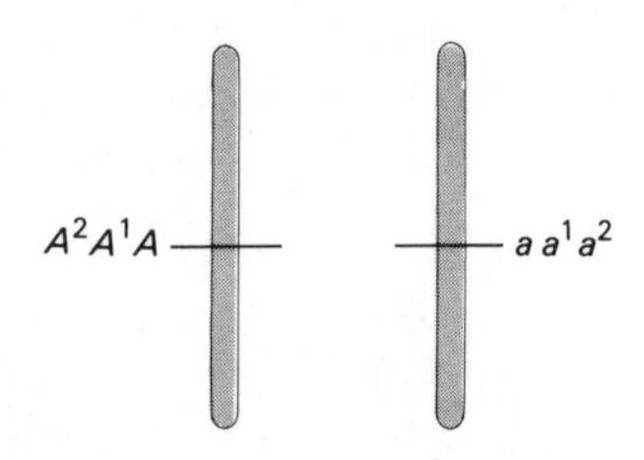

Figure 13. Multiple alleles. *A* and *a* are allelic genes occupying identical positions on homologous chromosomes. A^1A^2 and a^1a^2 are mutant forms of *A* and *a* (multiple alleles). Only one member of each series is present in any diploid individual but they can occur in any combination, e.g. Aa^1, A^1a^2, etc.

Studies at molecular level have shown that genes must represent groups or even portions of DNA molecules, i.e. numbers of atom-groups each consisting of a sugar, phosphate, and base linked together (nucleotide). How many nucleotides are needed to constitute a gene is still not certain, but in some instances at least it seems likely that they are one and the same. Similarly, as a result of extensive work on micro-organisms, present evidence points to the unit of mutation (sometimes known as a *muton*) being a single nucleotide. Although we are still far from understanding the biochemical nature of gene mutation, it now seems certain that this must involve an upset in the normal sequence of bases along the length of a DNA helix. Suffice it to add here that detectable mutations due to chromosome abnormalities and gene changes are rare, and appear to occur at a particular locus with a characteristic frequency. An approximate estimate of the average mutation-rate in living organisms due to all causes is about one in a million individuals, but the range of variation is considerable (see also page 137).

Sometimes the frequency is much higher, as in the gene causing haemophilia in man. This is a rare physiological condition in which the clotting mechanism of the blood is incompletely developed so that even slight wounds may result in excessive bleeding and be highly dangerous. Haemophiliac women are unknown for reasons explained shortly, while men seldom live beyond the age of twenty-five and therefore hardly ever produce children. Haemophilia is a hereditary condition and under the control of a sex-linked recessive gene (*h*)—that is to say, it is carried on the X chromosome. The most usual marriage by which the complaint can be transmitted to the next generation is between a normal man and a woman who is heterozygous for *h*. This means that in each of her cells, one X chromosome will carry *h* and the other *H*, its dominant allele. Thus, although normal herself, she will be able to transmit haemophilia since half her eggs will carry the *h* gene (see Figure 14). Haldane (1935)

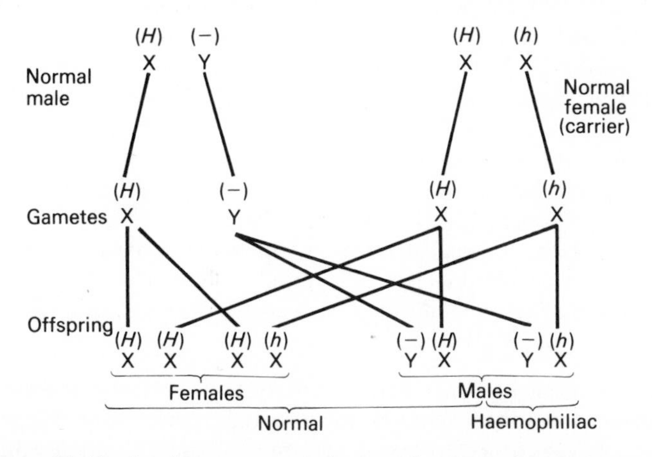

Figure 14. Marriage between a normal man and a woman heterozygous for haemophilia. All the female descendants are normal, but half are heterozygous and therefore carriers. Half the males are haemophiliacs

estimated the proportion of haemophiliacs per million male births in the Greater London area to be between 35 and 175. Since one-third of the X chromosomes in human beings occur in men (the other two-thirds being in women), it follows that one-third of haemophilia genes in a population (those in men) will almost certainly be eliminated at each generation. It is therefore possible to calculate the rate of replacement necessary to maintain a constant level, which is estimated as 1 in 50 000 individuals, and therefore represents the mutation rate. Owing to the comparative rarity of gene-mutations, such calculations are only possible at the highest rates.

Biochemical studies have shown that while the structure of the DNA molecule is complex, it is also remarkably stable. It is not surprising, therefore, that observed gene-mutations are so infrequent. However, as H. J. Muller first showed in 1927, their rate of occurrence can be greatly increased (up to 150 times) by the action of X-rays. More recently, similar effects have been demonstrated with a wide variety of so-called mutagens, particularly all types of short-wave radiation, where the incidence of induced mutation is found to be proportional to the dosage. Cosmic radiation in the atmosphere is, however, insufficient to account for more than a minute proportion of the mutations occurring in plants and animals under natural conditions.

All mutant genes exhibit certain characteristic features in common:

(i) Although confined to a comparatively small and limited range of possible changes, successive mutations at any one point show no evidence of a directional trend.
(ii) Mutations are restricted to specific loci; neighbouring genes and alleles are quite unaffected.
(iii) With rare exceptions, artificial agencies are not selective in their effect; that is to say, a given type of radiation does not induce mutation at a specific locus or of a particular kind.
(iv) Reverse mutation can occur; in other words, the same agencies can bring about a change from the 'abnormal' back to the 'normal' gene condition as are capable of inducing mutation in the opposite direction.

It is important to realize that changes of the kind just discussed are not confined to the nuclei of the germ cells, but may also occur in those of the body cells as well (*somatic mutation*). This has been demonstrated in both animals and plants, but from an evolutionary standpoint it is important chiefly in the latter, being the cause of *chimaeras*. In most plants the germ cells and somatic cells are intimately associated; indeed, the one is generally derived from the other. Hence it is quite possible for the results of somatic mutation to be passed on to the gametes and to become incorporated in the hereditary constitution of the offspring.

Contribution of mutation to variation

We have seen earlier (page 31) how the process of chromosome multiplication results in the formation of polyploids. These can be of two kinds. They may arise by the simple process of doubling, trebling and so forth; the chromosomes themselves remain unaltered, and hence pairing in the normal somatic condition is still possible.

Thus,

Parents	*AA* and *AA*		
Gametes	*A* × *A*		
Offspring	*AA*	(doubling) →	*AAAA*
Gametes	*A*		*AA*

Such forms are known as *autopolyploids* and are generally fertile with one another but not with other members of their stock having different chromosome numbers. They have proved to be valuable in plant breeding, where it has been found that the rate of mutation can be greatly accelerated by treatment with colchicine ($C_{22}H_{25}O_6N$), an extract from the seeds and corms of the meadow saffron (*Colchicum autumnale*), sometimes incorrectly called the autumn crocus. Its effect on the nucleus is to inhibit the appearance of the spindle at mitosis: thus, in the formation of the two daughter nuclei, the chromatid pairs fail to separate at anaphase and the number of resulting chromosomes is doubled.

Autopolyploids have been utilized commercially in many valuable crop plants. For instance, the triploid sugar beet grows more vigorously and gives a better yield than the normal diploid. It is unfortunately self-sterile, so seeds have to be produced by crossing tetraploid and diploid plants.

Under wild conditions autopolyploidy is well known, and in some instances it has been shown to have evolutionary value. Thus Hancock demonstrated three distinct chromosome races of the marsh bedstraw (*Galium palustre*). In Oxford the diploid ($2n = 24$) is found in damp places which dry out in summer, in Devon the tetraploid ($2n = 48$) occurs in wetter habitats which remain moist throughout the year, while the octoploid ($2n = 96$) is distributed round Oxford also in permanently wet sites. It is further distinguished by extensive vegetative reproduction by means of creeping shoots and root formation at the nodes.

The situation in *Galium* illustrates what appears to be a tendency among wild polyploid plants, namely, that the higher the chromosome number the greater the ability of the plant to colonize damp habitats. Indeed, in some species the external appearance of the plant varies so little that it is only by their ecology that the various chromosome races can be distinguished in the field.

It has been shown that the higher chromosome numbers in plants are by no means always multiples of one another and that differences between them are sometimes quite small. Such individuals are known as *aneuploids.* Lövkvist detected their occurrence in Sweden among plants of the lady's smock (*Cardamine pratensis*), which, like *Galium palustre,* shows a strong tendency for distribution to be influenced by the chromosome number. Those with the lowest number ($2n = 30$) were found in the higher dry localities, intermediate races ($2n = 56$-68) occurred lower down, while only the highest numbers were obtained from plants growing in or near water ($2n = 72$ and 76). Preliminary studies suggest that a similar situation exists in Great Britain.

A more widespread investigation by Yates and Brittan of the clover *Trifolium subterraneum* showed that the commonest form, collected from such diverse regions as England, Australia, Malta, and Portugal, has a diploid number of 16, while that occurring in Israel has only 12 chromosomes. The two races show no obvious external differences other than in their habit of growth, the Israeli plants being of a more spindly appearance. Attempts at hybridization have been unsuccessful and there is thus some justification for regarding the two forms as distinct species.

The second type of polyploidy arises in quite a different way. Suppose we attempt to cross two species belonging to the same genus, say a potato and a tomato (genus *Solanum*). If offspring appear at all, they will almost certainly be sterile. This is because the two sets of parental chromosomes are not homologous, and even if they happen to have the same number their constitution is different. They are thus incapable of pairing and normal meiosis resulting in the formation of gametes is impossible. But if by some irregularity of cell division the hybrid chromosome number should become doubled, the plant will then behave as a normal diploid and probably be fertile.

Polyploids of this kind, resulting from chromosome doubling following hybridization, are known as *allopolyploids.* The steps in their formation (e.g. in an allotetraploid) can be represented as follows:

Parents	AA and BB		
Gametes	$A \times B$		
Offspring (hybrid)	AB	$\xrightarrow{\text{(doubling)}}$	$AABB$
Gametes	–		AB
	(sterile)		(fertile)

Like autopolyploids, their rate of occurrence can be greatly increased by treatment with colchicine and short-wave radiations. Allopolyploids have been widely used in plant breeding, particularly in the production of new cereals. For instance, the cross between wheat (*Triticum*, $2n = 42$) and rye (*Secale*, $2n = 14$) gives rye-wheat (*Triticale*, $2n = 56$), an allotetraploid. Unfortunately, most strains of rye are habitually cross-fertilizing, unlike wheat and rye-wheat, which are self-fertilizing. This accounted at one time for a gradual degeneration of the rye component of the early rye-wheats and the high incidence of sterility– no doubt due to the effects of recessive genes reaching the homozygous condition as a result of continued inbreeding. The problem was eventually overcome by developing new strains of rye which were self-fertilizing, with the result that the proportion of seed set has now been increased from 50 per cent to 90 per cent of the yield obtained from standard wheat varieties. Flour derived from rye-wheat makes better bread than rye, and the plant is able to replace rye in sandy soils which are unsuitable for wheat cultivation.

Under wild conditions, a classical example of the success of an allotetraploid is provided by the cordgrass, *Spartina* (Figure 15). Our native species is *S. maritima* ($2N = 56$) and the first report of its hybridization dates from 1878 in Southampton Water. This was with the American species *S. alterniflora* ($2N = 70$), the result being a sterile hybrid *S. townsendii* ($2N = 63$) which can only reproduce vegetatively and still survives in some places. Chromosome doubling has, however, converted *S. townsendii* into a new and fertile form, ecologically superior to either of its parents and completely sterile with both of them. As is to be expected, the most usual chromosome number of the allotetraploid (sometimes still referred to as *S. townsendii* but, in fact, unnamed) is $2N = 126$, but plants with $2N = 120$ and 124 also occur. The new species has spread during recent years in a remarkable way along the south coast of England and has now achieved pest proportions, blocking large areas of Poole harbour and stretches of Southampton Water. It is one of the most successful colonists of tidal mud flats and has shown itself to be better adapted to this habitat than the native species. If grown in the right place, it can be of great value as a stabilizer

Figure 15. The cordgrass (*Spartina*), colonizing part of the estuary of the Beaulieu river, Hampshire

of ground in the reclamation of land from the sea. Incidentally, it is also used extensively in the manufacture of paper.

In plants capable of rapid vegetative multiplication allopolyploid sterility need not necessarily provide a complete check to successful colonization. Stebbins and Walters investigated the widespread occurrence of hybrids between various allopolyploid species of grasses of the genus *Elymus,* notably that between *E. condensatus* and *E. triticoides,* which is quite common in Central California. This is distinguishable from both parents by a number of structural features such as the type of growth and the shape and texture of its leaves. It is almost always sterile. The three forms have an interesting distribution which is clearly related to the peculiarities of each (Table 4):

Table 4. Distribution of three forms of the grass *Elymus*

Species	Distribution	Ecology
E. condensatus	Coastal regions	Brushwood slopes and open fields
E. triticoides	Inland; low-lying ground	Alkaline soils preferred
Hybrid (vegetative)	Inland	Grassy slopes; edges of fields, orchards, roadsides

The influence of man in burning grassland, ploughing soil, and constructing roads appears greatly to have aided the spread of the hybrid by means of rhizomes. Thus, in spite of its inability to reproduce sexually, the hybrid *Elymus* has been extremely successful on account of its vigour of growth, longevity, and ability to propagate vegetatively at a great rate.

The various types of chromosome fragmentation already summarized produce marked hereditary effects. Such major changes must cause an upset in the delicately balanced genotype, and it is not surprising that those mutations obtained under experimental conditions have nearly always been harmful to their possessors. Many instances have been reported in the breeding of *Drosophila,* generally following treatment with X-rays. Modern cytological methods have shown that numerous changes at one time attributed to the action of single genes are, in fact, due to the transfer of blocks of chromatin with their genetic complement from one position to another (*position effect*). The presence of such gene-groups in the homozygous condition is often lethal, and even as heterozygotes the resulting phenotypes are frequently less viable than the normal ones. Besides their physiological effects, mutant genes have been shown to produce many trivial structural differences. In *Drosophila* they include variations in eye colour, the number of bristles on the thorax, and the shape of the wings; features of no apparent significance under wild conditions. It thus seems unlikely that chromosome fragmentation plays any important part in initiating characteristics of potential value in evolution.

3

Variation in Populations

Some mechanism for the perpetuation of mutant genes is essential as a basis of genetic variation, indeed, without it our whole concept of particulate inheritance would be meaningless. Realization of this has prompted widespread experimental studies during recent years, and the results obtained have certain fundamental features in common. Of the many mutants detected in the laboratory, all are either recessives or 'semi-dominants', and the majority cause harmful physiological effects. Hardly any have ever been observed which could possibly be beneficial to an organism under wild conditions.

At first sight such conclusions would seem to upset our existing ideas, but certain precautions must be observed in interpreting them. In the first place it is important to remember that available experimental techniques enable us to detect only relatively large phenotypic changes. As we have seen when considering chromosome fragmentation, these are just the kind we would expect to cause an upset in the delicately balanced gene-complex and thus to be deleterious. Moreover, it is unlikely that the mutations studied are new; most of them have no doubt taken place many times before in the history of the species concerned. In *Drosophila,* for instance, the same mutation is known to have occurred more than thirty times at the locus of the gene causing white eyes.

Any mutant which endows its possessor with a better chance of survival in nature will naturally tend to become incorporated quickly in the gene-complex and be perpetuated by breeding. We would therefore expect that in the course of time the great majority of beneficial variations would have been selected positively and preserved. Such mutations would thus become progressively less apparent for study. Fisher estimated on general grounds that they might be expected to occur perhaps once in 10^9 individuals.

A striking example of the spread of an advantageous gene has occurred among certain species of moths inhabiting the industrial areas of northern England. One of these is the scalloped hazel *Gonodontis bidentata* (Figure 16). A century ago only the normal pale-coloured

Figure 16. Typical form (*left*) and melanic form (*right*) of the scalloped hazel moth, *Gonodontis bidentata* (x 1)

insect was known, but latterly this has been almost completely replaced in certain parts of the 'black country' by the dark form *nigra.* Its black appearance is due to the formation of the pigment melanin, an oxidation product of the amino-acid tyrosine. No such change has taken place in other parts of Britain where the pale form still remains predominant.

The distribution of *nigra* is of some interest, for while it has almost completely replaced the normal form in the area of Manchester, it has not yet established itself in Birmingham only eighty miles away, where a dark form is to be found exhibiting a wide range of variation and under polygenic control. The fact that the insect is a relatively reluctant flier may well account, at least to some extent, for its peculiar distribution.

As Ford pointed out, more than eighty instances of industrial melanism are now known among different species of moths in Britain, and large numbers have been recorded from other countries as well. Together, they constitute by far the most striking example of micro-evolution ever witnessed by man.

To return to *Gonodontis bidentata,* several important generalizations follow from a consideration of the evolutionary success of *nigra*:

(i) The genetic difference between the two forms is found to be a single allele, the melanic being *dominant* to the normal. Now, on our previous hypothesis, it is likely that this mutation has occurred many times before in the history of the species, but in the absence of a 'black country' it was of no adaptive value. Indeed, it must have been a great disadvantage, for melanics are extremely conspicuous against a natural woodland background, and it is therefore unlikely that many of them would survive being eaten by birds for long enough to breed. It is also equally improbable that the mutant gene, when it first appeared, was a true dominant. The activities of man have thus provided the circumstances necessary for two important and closely related changes in *Gonodontis bidentata*; the rapid spread of the new phenotype and its establishment in less than a century, and the speedy attainment of dominance by the gene concerned. This serves as a good illustration of the principle already outlined (page 23), namely, that it is not the genes themselves which are subject to selection but rather the genotypes in which they act. If, as in this instance, a gene produces a highly beneficial effect in a particular internal and external environment, those individuals will have the best chance of survival in which the mutant exerts its greatest influence. Selected continually in this way, those gene-complexes will tend to survive in which the new gene attains its maximum expression. In other words, it will soon become dominant to its corresponding alleles. Reasoning along these lines throws an interesting light on the quantitative nature of dominance and recessiveness.

(ii) No doubt the dark colour of the melanic moth provides concealment from birds and other predators in soot-blackened surroundings. But there is evidence that this is not the only advantage of *nigra*. Under experimental conditions it has been shown that the dark form can emerge at a lower temperature than the normal one, thus indicating a superior physiological hardiness as well. In another species, the mottled beauty (*Cleora repandata*), which also exhibits industrial melanism, the gene causing blackening in the adult is a benefit to the *larvae* as well, enabling them the better to withstand starvation, though it does not affect their colour. Here is yet another example of the physiological action of genes which, incidentally, draws attention to the fact that they often have multiple phenotypic effects, some of which may be advantageous and others disadvantageous in a particular set of circumstances.

(iii) A similar situation is also known in Germany where the melanic form of Geometrid moth, the brindled white-spot, *Ectropis extersaria*, has established itself in the industrial areas. In England the moth is locally common in the south but *not* in the manufacturing districts, although the black form, *cornelseni*, has been found there as a rare variety. Two important deductions can be made here: that industrial conditions are not necessary for the *production* of melanics, and that the black form is available for use by this species should conditions permit. Presumably, the reason why it has not spread is because the distribution of the moth is confined to rural areas.

Environmental changes of the magnitude and speed just described seldom occur. Hence it will be a rare thing for a gene causing such marked effects to spread through a population in this manner. Micro-evolutionary changes tend to be slow because:

(i) The circumstances in which animals and plants live under wild conditions are not normally subject to violent fluctuations.

(ii) The effects of mutation must usually be small if an upset of the gene-complex with its consequent deleterious results is to be avoided. The potential rate of evolution will thus tend to be greater in big populations than in small ones. For large numbers not only provide better opportunities for the spread and establishment of genes having small but beneficial effects, but they can also hold more genes in reserve whose influence in *existing conditions* is minimal. In a changing environment they may well prove of increased adaptive value.

It might be argued that if most observed mutations are harmful the normal forms of wild species must have been derived by some independent means. Strong evidence against this view is provided by the existence of *reverse mutation*—the ability of recessive genes to mutate back to the normal allele from which they originated. Thus Demerec showed that the gene causing miniature wings in *Drosophila virilis* often reverts to the wild form, the frequency and type of mutation (germinal or somatic) being partly controlled by other factors. Of particular importance is the fact that these changes can be induced by X-rays, since it has been suggested that mutant recessives obtained by such means under experimental conditions are abnormalities resulting from unusual treatment. This view is no longer tenable.

Domesticated species provide a special problem in the interpretation of mutation. We have seen how mutant genes observed in the laboratory are nearly always recessive or semi-dominants and usually harmful. Yet in stock breeding healthy varieties are well known to arise spontaneously. Sometimes the genes concerned are recessive and become apparent only as a result of inbreeding, but on occasions they can exert some effect in the heterozygous state and thus behave as partial dominants. It is difficult to judge the possible value of the effects of such genes under wild conditions. Suffice it to say that few would appear to have any use whatever, and most of them would probably be eliminated by natural selection. The characteristics produced by the 'dominants' are of just the kind to be valued by breeders as providing curios or increasing the yield of some commodity for human use, such as milk, meat or wool. Under conditions of domestication man will tend to select those gene-complexes in which the expression of a desirable character attains its maximum effect. Thus dominance of a gene can be rapidly attained; indeed, selection will be so severely in its favour that this can occur far more rapidly than would ever be possible under wild conditions. No doubt this accounts for the fact that some highly prized genes have actually become complete dominants and exert their full effects as heterozygotes.

Among poultry, the factor causing white plumage in the White Leghorn is an example of a gene which has become dominant in the gene-complex of the domesticated bird. When introduced into the genotype of the wild jungle fowl, the ancestor of modern poultry, the presence of the gene in the heterozygous state is still detectable but its phenotypic effect in producing white feathers is much reduced. In other words, by returning the gene to the wild genotype from which it originated, thereby reducing its degree of dominance, a reversal of the supposed evolutionary process can be brought about artificially.

Polymorphism

We have so far been concerned with instances in which a dominant gene or group of genes exerts effects which are a definite advantage and where the recessive alleles seldom have a chance of expressing themselves. On the rare occasions when they do so, the resulting phenotype will almost always be eliminated under wild conditions. But it sometimes happens that a species exists in two (or more) quite distinct forms which occur together in the same habitat and are controlled by a simple genetic mechanism. On occasions the rare variant may be so infrequent that its occurrence can be attributed to mutation, a typical example being albinism (total absence of pigment) in birds such as blackbirds. In general, animals are extremely sensitive to 'abnormalities' in their midst, and it is unlikely that an albino would succeed in finding a mate and therefore in transmitting the mutant gene to the next generation. An altogether different situation occurs when two or more forms of the same species occur together, the rarer of them at a frequency *above* that of recurrent mutation. This is known as *genetic polymorphism*.

Polymorphic forms tend to be extremely sensitive to the influence of natural selection; indeed, they themselves are products of it. As we have already seen in the scalloped hazel moth, *Gonodontis bidentata*, a situation can arise in which changes in the environment may cause one form to succeed another, the less advantageous variety becoming progressively rarer. This is what happened when the normal form of the moth was being replaced by the melanic in certain industrial areas. While the process was going on, the proportions were constantly changing in favour of the new variety until it finally achieved predominance. Such a situation is called *transient polymorphism*.

A second and more complex condition arises when two or more forms are maintained at some *fixed* level on account of the advantages they both confer in a particular situation. This is *balanced polymorphism*, but in achieving equilibrium it must pass through a transient phase. The existence of balanced polymorphism in animals and plants has been known for a long time and several examples were described by Darwin in his *Forms of Flowers* (1877). One of these is the variation in the style-length (heterostyly) of the common primrose, *Primula vulgaris*. The flowers generally occur in two forms (dimorphism). In the 'pin-eyed' flower the style is long with a stigma near the mouth of the corolla tube, and the five anthers are situated half-way down it. The 'thrum-eyed' form has a short style, the stigma being about half-way along the corolla, and the anthers round the mouth (Figure17). The flowers are insect-pollinated and pollen naturally tends to be transferred from the anthers of one type to the stigma of the other. The dimorphic flowers thus provide a simple structural mechanism for encouraging cross-pollination and therefore the maintenance of variation. The results obtained by Darwin when studying the various possible crosses are summarized in Table 5.

Table 5. Darwin's crosses of the primrose (*Primula vulgaris*)

Cross	Flowers pollinated	Capsules set	Average seeds per flower pollinated
Normal			
thrum x pin	8	7	56.9
pin x thrum	12	11	61.3
Abnormal			
thrum x thrum	18	7	7.3
pin x pin	21	14	34.8

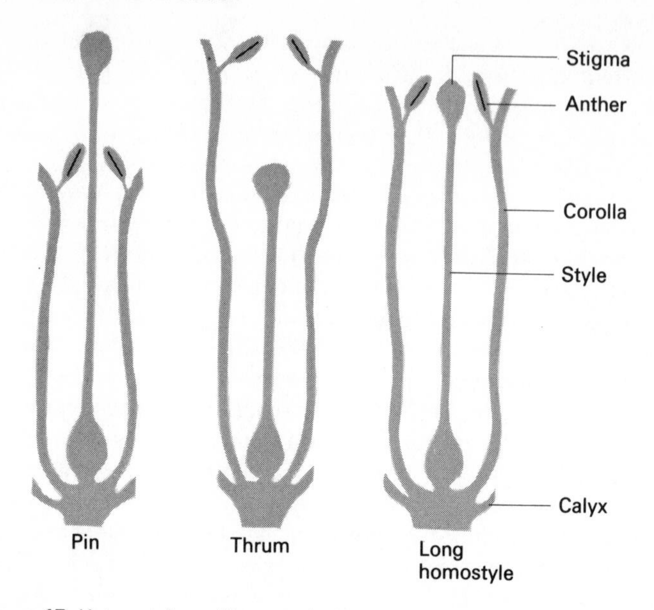

Figure 17. Heterostyly and homostyly in the common primrose, *Primula vulgaris*

There is a physiological difference here as well as a genetic one, for the fertility of normal crosses is obviously greater than that of abnormal ones. This is due to the different growth rate of the pollen tubes. If in any cross a mixture of 'normal' and 'abnormal' pollen is placed on a stigma, the 'normal' will always grow the faster and achieve fertilization. The results of the three possible crosses are as follows:

Pin × pin	All Pin
Thrum × thrum	3 Thrum : 1 Pin
Pin × thrum	1 Thrum : 1 Pin

The genetic mechanism thus consists of a single pair of alleles (*S* and *s*), thrum being dominant. In nature, all thrum plants are heterozygous (*Ss*), thus normal pollination results in a back-cross from which 50 per cent of each form is produced.

But there is a further problem to be considered. If the growth rates of the haploid pollen grains are studied when applied to different stigmas, we find that all those derived from *Ss* plants behave in a *thrum* way irrespective of whether they carry *S* or *s*. All *s* pollen from *ss* plants behaves as pin. Thus there are actually two 'switch' mechanisms involved in determining compatibility:

(i) Germinal, acting only through the pollen grain chromosomes.
(ii) Cytoplasmic, the nature of the pollen being determined by the cytoplasm of the diploid parent.

Counts of wild primrose populations in Dorset and Surrey have shown the two forms to exist there in almost exactly equal numbers. In Somerset, however, the situation is quite different. Turrill gives the figures in Table 6 for five colonies.

Table 6. Distribution of pin, thrum, and long-homostyle plants in five primrose colonies. (After Turrill.)

Colony	Pin	Thrum	Long homostyle (see Figure 17)
1	102	11	210
2	145	15	468
3	152	103	177
4	46	75	7
5	41	40	2

The proportions of pin and thrum obviously vary greatly in the five habitats; the reasons are unknown. Suffice it to say now that situations of this kind are common among polymorphic species and indicate the varying action of natural selection in different circumstances. This will be discussed further in the next chapter.

It will be noted that a new form, long homostyle, has also appeared. This has flowers with stamens round the mouth of the corolla (thrum condition) and a stigma in the pin condition, and is self-fertilizing (Figure 17). Mather and de Winton put forward a convincing explanation of its significance. They pointed out that the genetic mechanism involved in normal heterostyly, as we have already seen, controls three characteristics: (i) the length of the stamens, (ii) the length of the style, (iii) physiological compatibility. It is therefore quite possible to imagine another allele occurring at the *S* locus which might fail to alter the relative positions of the anthers or stigma while readjusting the illegitimacy behaviour. Thus homostyly would become superimposed on a previous heterostyle condition, with consequent inbreeding instead of outbreeding. The new condition might well have certain evolutionary advantages depending on the status of the population in question. In some circumstances it might be desirable for a species to remain highly variable and therefore heterostylic. Such a situation appears to exist in most of the primrose populations so far studied. On the other hand, occasions must sometimes arise when it is to the advantage of a well-adapted form to maintain its adjustment to the optimum conditions and to vary as little as possible. In such circumstances natural selection would favour the homostyle form and the process of inbreeding that accompanies it. This may well be the explanation of mixed heterostyle and homostyle communities.

It was originally believed that homostyle primroses were capable of reproducing only by self-fertilization, but Bodmer has drawn attention to a fallacy in this assumption. If it were so, the chances of producing seed in homostyle flowers would obviously be far greater than in heterostyles requiring cross-fertilization and there would be no reason why, once established in a population, homostyles should not sweep through it and soon become the predominant form. All the available evidence suggests that such changes do not, in fact, occur. The discovery that under natural conditions cross-fertilization occurs in a high proportion of homostyles points to the existence of an effective controlling mechanism. For if, as in the case of heterostyly, outcrossing among homostyles is associated with differing degrees of compatibility, it is not difficult to envisage a situation in which the amount of crossing might provide a delicate control over the production of homostyles. Indeed, evidence from counts suggests that in some populations homostyle plants may already have reached a threshold value or even be declining in numbers, probably as a result of such a selective mechanism.

It is interesting to note that in one British species, *Primula scotica,* an inhabitant of the north of Scotland, only homostyle flowers occur. On our present assumption this is unlikely to be a primitive condition, but rather a highly specialized inbreeding device superimposed on former heterostyly. Indeed, it now seems certain that *P. scotica* evolved from the closely related *P. farinosa,* itself a heterostyled form.

Continuous variation

While polymorphism is concerned with relatively large and clear-cut differences, the great majority of variation is of quite a different kind. We have seen how most gene changes must necessarily be small if they are to be beneficial, and should therefore expect comparatively minute differences among the organisms that carry them. Variation in such characteristics as size, weight, colour, and shape are often difficult and laborious to assess with any accuracy. Yet such studies are well worth making, for it is only by obtaining quantitative information that differences due to the action of selection can be determined. Man himself provides some excellent examples, thanks to the greatly increased efficiency with which his biometrical records are now kept. Thus, Karn and Penrose studied the birth weights of some 13 730 children born in a London hospital between 1935 and 1946, and their distribution is shown in Figure 18. Data are also available on the survival

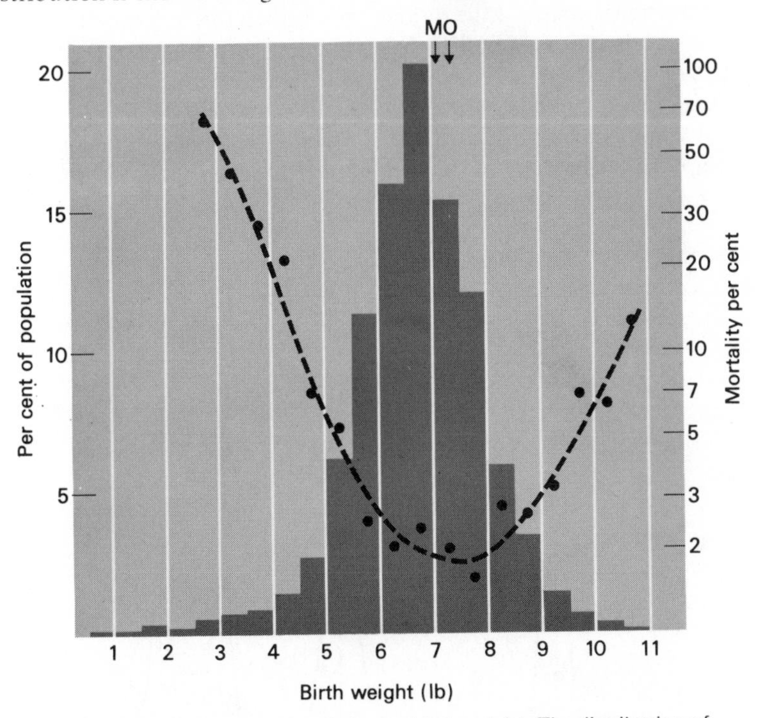

Figure 18. Child mortality in relation to birth weight. The distribution of birth weight among 13 730 children and the rates of early mortality in the various birth weight classes. The shaded diagram shows the proportions of the population falling into the various classes in respect of birth weight. The broken line is the curve of mortality in relation to birth weight, the values actually observed being represented by the points to which the curve is an approximation. The percentage mortality is set out as a logarithmic scale for ease of representation. *M* marks the mean birth weight and *O* the birth weight associated with the lowest mortality and hence to be regarded as the optimum. (After Mather.)

of the babies, which show that 614 (4.47 per cent) were either still-born or died during the four weeks after birth. The relationship between mortality and birth weight is plotted for each weight class, mortality being assessed as the percentage of children failing to survive to four weeks. Evidently, the optimum birth weight for survival (*O*) is very close to the mean (*M*), while on either side the expectation of survival declines rapidly, reaching a minimum at the two extremes. When considering data of this kind, it is important to bear in mind that at least part of the variation will be attributable to environmental causes, such as the nutritional status of the mother, and is therefore not inherited. In this instance, it was estimated that rather less than 50 per cent of the variation in birth weight must have been due to genetic causes, the remainder being non-hereditary.

Situations such as that outlined above are sometimes quoted as examples of *stabilizing selection* (Figure 19), the optimum quantity

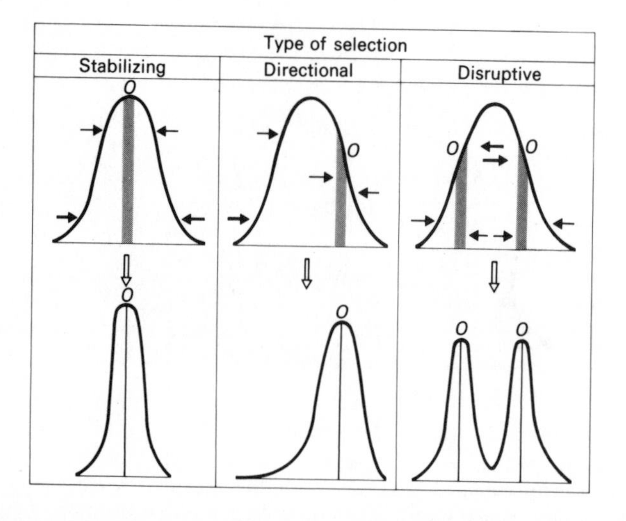

Figure 19. Three types of selection. The upper diagrams show the optimum phenotypes selected (*O*), and the direction and thickness of the arrows indicate the direction and magnitude of selection. The lower figures show the results of selection after a partial response. In stabilizing selection the optimum is the average expression of a characteristic; in directional selection it is either greater or less than the average; in disruptive selection there are two optima. (After Lewis and John.)

selected representing the mean value, or nearly so, and the two extremes tending to be at a disadvantage. However, situations can arise in which circumstances may favour a form well removed from the mean, in which case selection is said to be *directional* (Figure 19). Thus the grayling butterfly, *Eumenis semele,* a common inhabitant of chalk downland and dry heaths, is distributed throughout the British Isles except in the extreme east. The average wing span of males is 48 mm and in females 52 mm, the limits of variation in each sex being 3-4 mm. A. J. Thompson has described a remarkable local race known as *thyone* which is found only in the area of Great Orme's Head, North Wales. Here, the average wing expanse is 41 mm in the male and 43 mm in the female. Nothing is known as yet of the significance of this dwarf variety nor why it should be able to maintain itself only in this remote locality. Presumably, small size and limited flying ability could be advantageous in exposed and windswept conditions. Ford has drawn attention to the fact that *thyone* emerges as an adult several weeks earlier than the normal form in the same vicinity. He suggests that the dwarfs may carry a gene or genes which speed up development in the larval stage. Thus pupation would take place before the larvae had had time to attain their maximum growth, so giving rise to miniature adults.

A third situation may arise, *disruptive selection* (Figure 19), where there is more than one optimum expression of a particular variant maintained by directional or stabilizing selection, or by a combination of both. This possibility may be particularly useful in helping to explain the problems posed by *clines* (gradients of variation in a species within a particular zone). Examples of these will be considered in Chapter 4.

When variation within an animal population fluctuates in response to changing circumstances, it does not necessarily follow that reactions by the two sexes will be the same; sometimes they may differ appreciably. This situation and its wider implications for the study of variation are well illustrated in the meadow brown butterfly, *Maniola jurtina,* in which the hind-wings have a row of black spots on their under-side varying in number from 0 to 5 (Figure 20). Ford and I chose to study this characteristic because, being quantitative, it was likely to be under polygenic control, and therefore to respond quickly to the effects of natural selection. Extensive samples across southern England westwards to the vicinity of the Devon-Cornwall border, over a period of twenty years, have revealed a constant and characteristic spot-distribution for each sex (Figure 21). While both types are unimodal, that of the male has the large mode at 2 spots and the female has a mode at 0. Evidence

Figure 20. Meadow brown butterfly, *Maniola jurtina,* at rest. *Left,* male showing the two-spot condition and *right,* female with no spots (x 1)

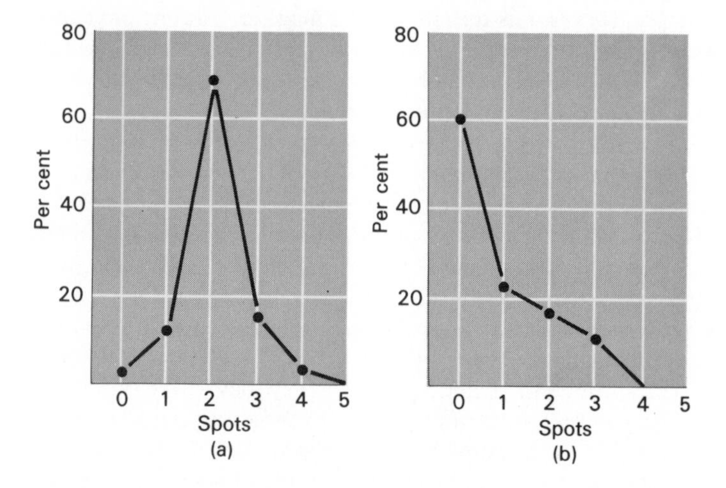

Figure 21. Spot-distribution in the meadow brown butterfly, *Maniola jurtina.* The Southern English and General European type: (a) males and (b) females

derived from a study of extensive collections in the Natural History Museum has shown that the Southern English spotting is but the western extremity of a far larger stabilization extending across central

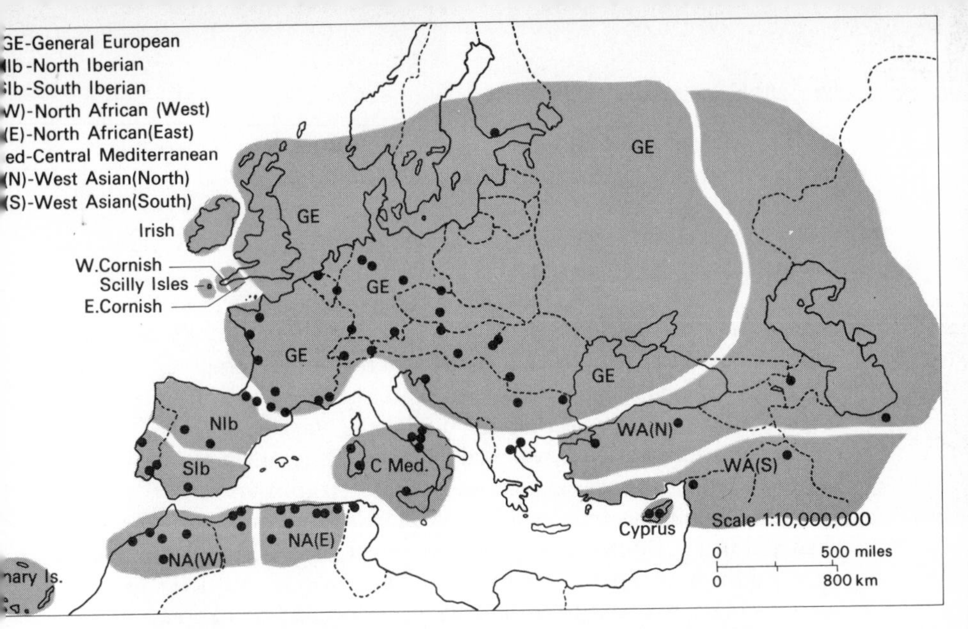

Figure 22. Spot-stabilization in the butterfly, *Maniola jurtina,* throughout its range. Dots denote areas where samples were obtained outside the British Isles.

Europe into Asia Minor–General European stabilization (Figure 22). Some evidence of its antiquity can be gauged from the fact that the earliest museum samples date from 1890.

The situation in *Maniola jurtina* outlined so far serves to highlight certain important biological principles which are worth summarizing:

(i) The range of spot-variation in the adult is strictly limited, and is the same in both sexes. Normally this is from 0-5 spots, although occasionally, a 6-spot individual is found. Evidently, the gene system involved is strongly homeostatic in its effects. At present we do not know exactly how much of this variation is genetic and how much environmental, although it is now clear (McWhirter 1969) that part, at least, is inherited.

(ii) From a careful study of many thousands of living insects in their natural environment, there is no evidence that the number of spots on the hind-wings is, in itself, of any survival value. For instance, they play no part in courtship or in disruptive camouflage. However, it now seems clear that the genetic mechanism controlling them has multiple effects, one of these being to determine the extent of the insect's liability to parasitization in the *larval* stage. The implications of this in relation to natural selection will be discussed further in the next chapter.

(iii) Perhaps the most striking feature of spotting is its remarkable *stability* over a wide geographical area, and for a long period of time. It is an axiom of ecology that, under natural conditions, the larger and more diverse a community, the more stable will it tend to become. Clearly, the larger the area, the greater the diversity of ecological niches that it can support. Should the staple diet of a particular animal species fail, the chances are that it will be able to find a suitable alternative. By contrast, small communities tend to be unstable and hence to respond rapidly to the effects of environmental change. As we shall see later, this principle is also well illustrated in certain populations of *Maniola.*

The spot-stabilization considered so far can be regarded as a basic pattern for the two sexes and represents the extent of *first-order* variation. Within this framework we might expect to find *second-order* changes, such as those occurring within a single season or between one year and the next. Spot-distribution does not provide a sufficiently sensitive index to quantify such differences, so a different parameter such as *spot-average* must be used instead. For any sample of *Maniola,* this is obtained by multiplying the class values by their frequencies, summating, and dividing by the total. Some typical results from a locality studied over a period of eight years are summarized in Table 7 and shown graphically in Figure 23.

Table 7. Summary of samples of *Maniola jurtina* from the Winchester area (1961-8) to show second-order variation (from *Heredity*).

Date	Sex	Spots 0	1	2	3	4	5	Total	Probability	Spot-average	Range of variation
June-July	Male	16	70	808	260	75	5	1234	$0.01>P>0.001$	2.26	2.12-2.41
August	Male	6	27	413	85	14	2	547	$0.3>P>0.2$	2.15	2.01-2.25
June-July	Female	492	329	209	94	10	–	1134	$0.05>P>0.02$	0.94	0.74-1.21
August	Female	505	198	104	22	2	–	831	<0.001	0.58	0.42-0.97

The picture that emerges from this study of the Winchester population of butterflies is one of extreme spot-stability, in so far as first-order variation is concerned. For eight seasons, the males maintained the characteristic high mode of 2 spots while the females were unimodal at 0. However, the population showed considerable second-order variation both within seasons (intra-seasonal) and between them (inter-seasonal).

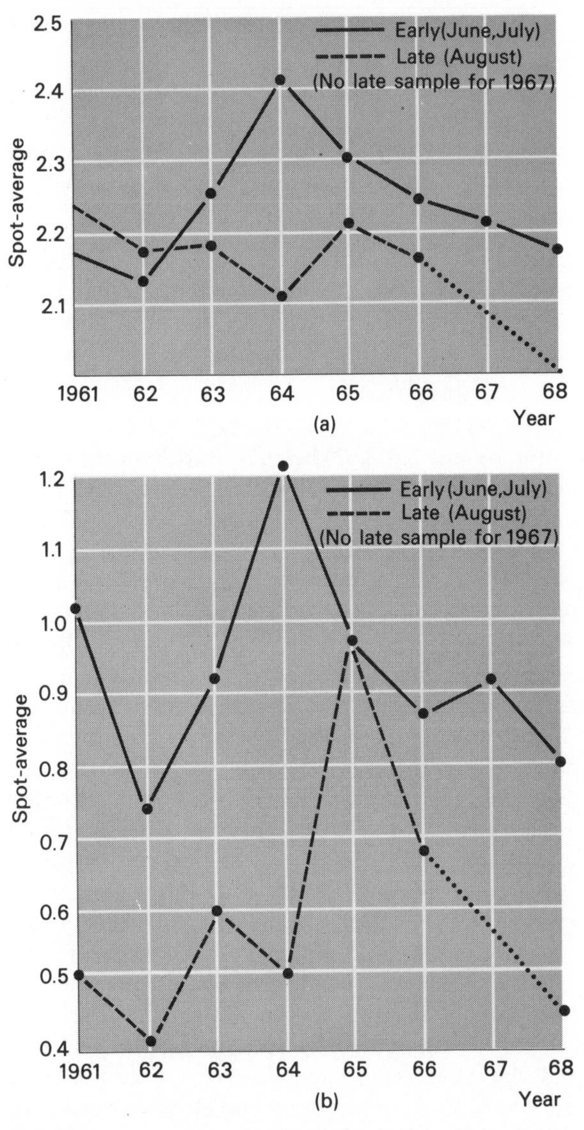

Figure 23. Intra- and inter-seasonal variation in the meadow brown butterfly, *Maniola jurtina,* in the Winchester area, 1961-68: (a) males and (b) females

The data in Table 3 and Figure 23 suggest the following general conclusions:

(i) The column for 'Probability' in Table 7 provides information on the homogeneity of each set of samples. The low values of P indicate a high level of intra-seasonal variation, which is expressed graphically in Figure 23.

(ii) Inter-seasonal fluctuations in the two sexes followed much the same pattern but were greater in the females than in the males (Figure 23). This accords well with studies of first-order variations in spot-distribution, where the female is also found to be the more variable sex.

(iii) Intra-seasonal variation followed the same pattern in both sexes, the spot-average (Table 7) being high early in the season and lower later on. Thus, spotted females comprised 57 per cent of the early samples but only 39 per cent of the late ones.

Summarizing the situation, we may say that while studies of first-order variation in *Maniola* reveal a remarkable stability, second-order variation is found to fluctuate inter-seasonally within a stabilized framework, and also intra-seasonally with a regular periodicity.

At present, we know little of the mechanisms controlling variation in *Maniola.* The fact that inter-seasonal variation is a feature of all meadow brown populations studied so far, suggests that the environment must exert an important influence. This hypothesis is difficult to test for a variety of reasons, not least because so little is known about the micro-climate of ecological habitats such as grass tufts, where the animal spends all but about a month of its life. That this environment differs greatly from the surrounding macro-climate is evidenced by the difficulty, so often experienced by collectors, of predicting the responses of larvae to particular weather conditions. On the other hand we have seen that such spot fluctuations as occur, do so with a distinct regularity, suggesting at least partial control by a genetic mechanism. As McWhirter has suggested, the difference between first- and second-order variation in an animal such as *Maniola jurtina,* may provide some measure of the antiquity of the genotypes involved. First-order spot-stability, which may well extend to other closely related genera within the Family Satyridae, could be controlled by relatively ancient gene systems (*palaeogenes*) producing phenotypes with reduced sensitivity to the fluctuating effects of this environment. Second-order variation resulting from a relatively high sensitivity to environmental change, may be governed by more recently evolved gene-systems (*neogenes*).

Sexual reproduction and variation

The previous discussion of the nature and magnitude of variation leads us to consider how variation is transmitted and perpetuated. Many living organisms are known to reproduce by some asexual means. Among plants vegetative reproduction is widespread, and apomixis (the setting of seed without fertilization) also occurs in a number of species. In animals fission and budding are often found among lower forms, while parthenogenesis occurs widely in insects and other groups.

Now it is a fact that many organisms which reproduce asexually also exhibit a sexual phase. The relationship between the two kinds of reproduction has been widely studied, an interesting example being the ciliate protozoan *Paramecium.* Like many simple animals and plants, *Paramecium* habitually reproduces by binary fission. But in most species a second type of reproduction (*conjugation*) has also been observed in which two individuals come together and fuse temporarily with their oral surfaces in contact. After a varying number of nuclear divisions, one of which is a reduction division, reciprocal exchange of chromatin takes place and the resulting diploid nuclei are those carried by the offspring. Although there is no visible differentiation of gametes, the process can clearly be regarded as a kind of sexual reproduction in which gene recombination doubtless occurs, so promoting variation. But this is not all. About once a month *Paramecium* undergoes a second and peculiar type of nuclear rearrangement called *autogamy.* The changes involved are similar to those in conjugation, but the animals do not pair. The meganucleus soon disappears and the micronucleus divides three times, one of the first divisions being meiotic. The eight nuclei so formed are thus all haploid. Seven of these degenerate and the remaining one divides again into two products which soon reunite.

As Beale has pointed out, autogamy thus represents the most extreme form of inbreeding in which the resulting animals become homozygous for all their genes in a single step. In Figure 24 the process is represented diagrammatically for a pair of chromosomes, *a* being a 'normal' gene carried on one of them, and *a′* its mutant allele on the other. It will be noticed that any individual starting in the heterozygous condition has a 50 per cent chance of becoming homozygous for the new gene.

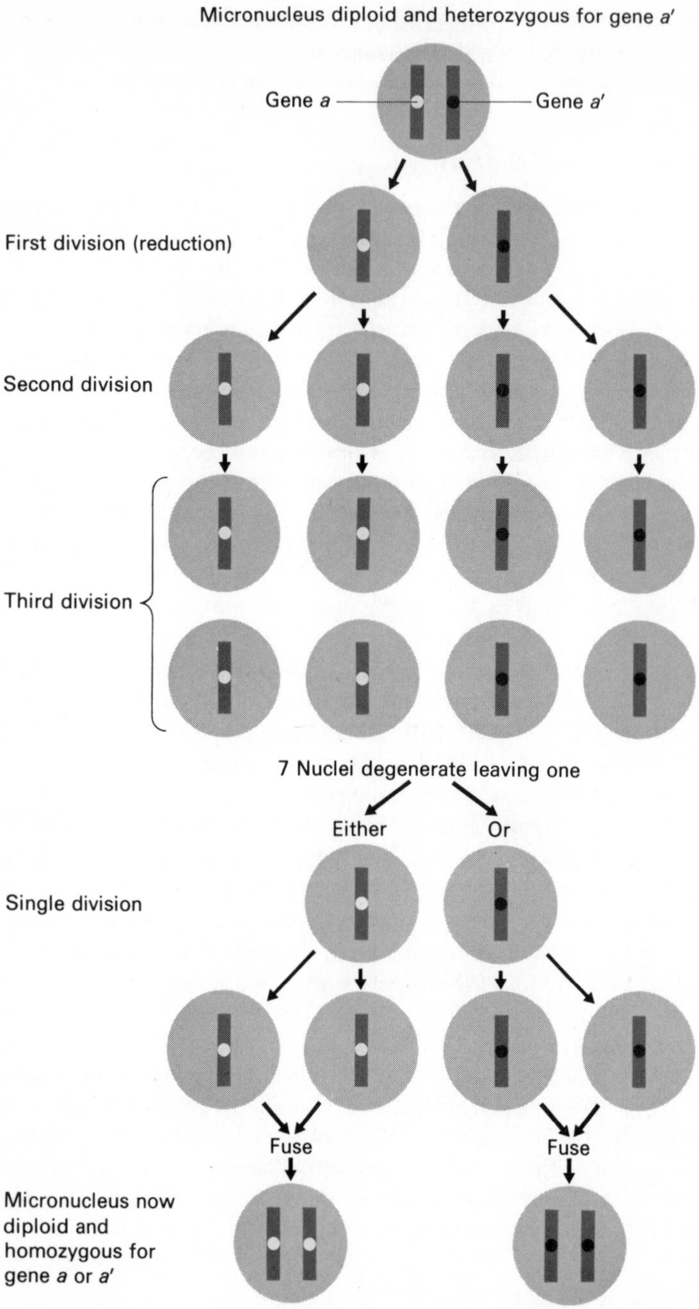

Micronucleus diploid and heterozygous for gene *a′*

Gene *a*

Gene *a′*

First division (reduction)

Second division

Third division

7 Nuclei degenerate leaving one

Either

Or

Single division

Fuse

Fuse

Micronucleus now diploid and homozygous for gene *a* or *a′*

Figure 24. The distribution of a pair of genes in autogamy of *Paramecium*

It used to be thought that cultures of *Paramecium* deteriorated if conjugation did not take place but Jennings showed that this is not so. In fact, *provided conditions remain favourable,* propagation by fission can continue indefinitely, but only if it is accompanied by autogamy at regular intervals. On genetic grounds this is rather strange, for after the first autogamy has taken place and the homozygous condition is achieved, no further nuclear change is possible. It seems likely, however, that there may be other benefits resulting from autogamy which concern the meganucleus. The reason for its large size is now known to be the greatly increased number of chromosome pairs that it contains (polyploid condition). Unlike the micronucleus, division of the meganucleus is amitotic; that is to say the chromosomes which it contains are distributed in the products entirely at random. One consequence of this process is believed to be a gradual genetic unbalance which results in a progressive decrease in viability and an increase in various structural abnormalities among the daughter *Paramecia.* In both autogamy and conjugation a new meganucleus is formed and the correct balance of chromosomes is thus restored, with a consequent improvement in the vigour of the stock.

The situation in *Paramecium* illustrates an important biological principle, namely, that most organisms which normally reproduce asexually have recourse to sexual reproduction when conditions become adverse. An advantage of this, which is common to all organisms that reproduce sexually, is the assurance of continued variation and with it a better chance of survival in changing circumstances.

But *Paramecium,* and possibly many other small organisms, have carried the reproductive process a stage further. For autogamy ensures that any genes having beneficial effects will rapidly achieve a homozygous condition and therefore exert their full effects in many members of the population. Similarly, disadvantageous recessives will be unmasked and exposed to the full force of natural selection. The process is thus of both evolutionary and eugenic value, quite apart from any other physiological advantages which it may confer.

4

Natural Selection and Adaptation

We have seen in the previous chapters how the forces of selection acting upon the enormous fund of variation resulting from Mendelian inheritance enable organisms to become adapted to the manifold conditions in which they live. But when we speak of *adaptation* in a particular animal or plant we often do so with considerable ambiguity. For instance, we look upon the thickened skin on the heel of the human foot as an obvious adaptation for easier walking, and so it is. At birth the skin in this region is thicker than in any other region of the body, long before the foot has ever made contact with the ground. Moreover, transplantation experiments show that if a piece of the sole of a guinea pig's foot is grafted on to the chest, it continues to behave exactly as before although relieved of all pressure, and forms the characteristic thick cuticle in its new position. Comparable are the callosities on the hands of a gardener, resulting from manual labour, but this kind of adaptation is obviously different for it does not appear unless the necessary external stimulus is applied.

Medawar has subjected such conditions to a critical analysis in which he points out that three types of adaptation in fact occur. Typical examples of each are:

A. The transparency of the epidermal cells and the dermal fibres that form the cornea of the eye.
B. The thickness of the epidermis on the heel of the foot, mentioned above.
C. The well-developed muscles of the arms of the blacksmith.

Class A adaptations are all innate and cannot be achieved by an individual as a result of the influence of the environment. No amount of use will convert an opaque epidermal layer into a transparent one. On the other hand, growing transplants of the corneal tissue of a rabbit maintain their transparency irrespective of the part of the body into which they are grafted. Class B adaptations are similarly inborn but are of a kind which could be acquired temporarily during a lifetime even if no genetic mechanism existed to control them. Structures of this sort which simulate others resulting from heredity, but are not themselves inherited, are sometimes referred to as *phenocopies*. Class C adaptations can be regarded purely as somatic modifications resulting from the effects of use or disuse; they are not inherited. It was with this latter group that the Lamarckians were intimately concerned, for implicit in their theory was the assumption that Class C adaptations could be converted into Class B or even, perhaps, into Class A. But this notion of inter-class transformation suffers from a serious logical limitation, for it does not explain in the last resort how the *ability* to achieve Class C adaptations was evolved.

One of the great merits of Darwinism is that it accounts for adaptation purely in terms of survival values, thus eliminating the necessity for postulating inter-class changes. Furthermore, its fundamental requirements, namely variation and natural selection, are well known and can be demonstrated. It is true of course, that much obscurity remains in the way of a logical Darwinian interpretation of many well-known evolutionary trends. Thus we know virtually nothing of the way in which hormone systems have evolved in plants and animals, and it is extremely difficult to picture the various transitional phases through which they must have passed. On the other hand, as we shall see in the following pages, we have much positive and precise evidence concerning the mode of action of natural selection in particular instances, which gives us good grounds for believing in its more widespread application.

But the important point to realize here is that, while the present limitations of Darwin's theory extend to all three kinds of adaptation, it is equally able to explain many instances of each kind of evolutionary change. Lamarckism, on the other hand, was quite incapable of accounting for the great bulk of Class A evolution. From such a discussion it can be seen that adaptations of Class A and C are by far the commonest, but that only the former are inherited and therefore of value in evolution. These will be our main concern when considering the mechanism of natural selection.

Natural selection and specialization

The ability of an animal or plant to colonize a new type of habitat and maintain itself there ultimately depends on two factors. First it must achieve a certain level of physical and physiological tolerance to enable it to gain a foothold in the new locality—i.e. it must attain some degree of *pre-adaptation.*

A second necessity is sufficient genetic variability to enable it to establish itself in the face of such selective agencies as the climate and competition from other animals and plants. This phase is sometimes known as *post-adaptation* and includes all subsequent responses by the organism to the constantly changing environment. Varying degrees of structural and physiological change may result. Some species seem remarkably tolerant of a wide range of conditions. Nematode worms, for instance, have successfully colonized such diverse habitats as vinegar vats, soil, nephridia of earthworms, and the alimentary canal of man without undergoing much structural modification in the process. Among parasitic forms, the genus *Heterodera* alone is said to colonize no less than 850 different species of wild and cultivated plants. On the other hand, the Cestodes (tapeworms), all of which are parasites, have undergone profound changes in order to become adapted to the diverse environments provided by their particular hosts. They are thus said to exhibit *specialization.* Nearly all of them inhabit the small intestine of vertebrates, feeding on various carbohydrates absorbed from their host's food and certain specific nitrogenous compounds derived from the intestinal wall. Most of them are confined to a restricted group of animals, or even a single species. Indeed, so specific have some of them become that the study of their structure and distribution (comparative parasitology) has thrown much new light on the evolutionary history of their hosts. For example, it has been shown by Baer that the African ostriches and South American rheas both harbour a species of tapeworm of the genus *Houttuynia* which is found in no other birds. The Australian cassowaries and emus, on the other hand, have tapeworms belonging to the widespread genus *Raillietina.* The resemblances between the parasites of the ostriches and rheas also extend to their Sclerostomes (Nematodes) and feather-mites as well, none of which resemble those found in other birds. There is thus strong evidence of a common origin, but one distinct from the ratite birds of Australia (cassowary and emu).

Excessive specialization of this kind is a sure prelude to extinction. Once such a course of 'over-adaptation' has been pursued, it can never be retraced, for in the process an organism must inevitably sacrifice its plasticity, and hence its ability to evolve further. This is particularly true of tapeworms, where habitual self-fertilization must bring about a large measure of homozygosity and a subsequent reduction in variation.

Biological history is full of instances in which varying degrees of specialization have served to lessen the power of organisms to evolve in a changing environment, and hence have been responsible for their eventual extinction. The condition already described in the natural colonies of *Primula vulgaris* (page 48) thus appears to be an ideal one in which the two mechanisms of heterostyly and homostyly exist together, the former promoting variability by out-breeding, the latter leading to a stabilized inbred form adjusted to the optimum conditions existing in a particular place.

The nature of selection

One of the limitations of our present theory of evolution is its frequent inability to account for the selective agents concerned in evolutionary changes.

When describing an environment in which animals and plants live, it is customary to separate the various factors concerned roughly into two groups, *climatic* and *biotic*. Climatic factors are those of a physical nature, such as temperature, light, and humidity. Biotic factors are less easy to define, being the outcome of the inherent peculiarities of the organisms themselves. They include such variables as food and living space.

Now a detailed examination of any community will soon reveal one important fact, namely, that all these influences generally interact in a most complex way to determine the relative survival of the various organisms. More often than not it is quite impossible to disentangle the exact part played by any one factor or group of factors. We must also remember that natural selection may operate differentially in the young and older phases of the life cycle (i.e. be *endocyclic*), a fact which greatly complicates its study. The various factors concerned in determining growth-rate, for instance, are still far from clear. We now know a good deal about the influence of climatic factors such as temperature and humidity on germination in plants, but we are still a long way from understanding precisely the circumstances in which weeds have managed to acquire their phenomenal powers of growth and multiplication.

In animals the operation of selection during early life may be more obscure still, as is shown by a study of pre-natal mortality in the hare. Of all the litters conceived, no less than 60 per cent are never born at all, while among those that survive a loss of between 9 and 10 per cent occurs during the course of development. The lost embryos are not expelled from the uterus as might be expected, but are absorbed within it by a process resembling phagocytosis. The likelihood of loss is found to be correlated with the initial size of the litter, and to be highest when the number of developing embryos is unusually large or small. The average number of eggs released from the ovary during each oestrous period is about five, and it is significant that the litters most likely to survive to birth are those which start with five or six young.

The cause of this remarkable process of elimination still remains obscure, but it may well be perpetuated by natural selection, for it seems likely that any mechanism which can adjust the number of embryos to harmonize with the physiology of the parent will tend to provide an optimum environment for development and hence be of survival value.

Brambell and Mills studied the embryology of the hare and have thrown some interesting light on the circumstances in which the death of the embryos occurs. At an early stage of development the yolk-sac, which connects with the embryonic gut by the umbilical cord, becomes filled with fluid. This is normally quite clear, but in embryos destined to die about the twelfth day the fluid becomes gelatinous and cloudy on the eighth or ninth. This is due to a kind of clotting and the production of the insoluble protein fibrin. Why some embryos should be affected and not others is unknown, nor is the relationship clear between the clotting of the fluid and the death of the embryo. It does, however, seem practically certain that the soluble precursor of fibrin passes into the yolk-sac from the blood of the mother. There is thus a possibility that the hare exhibits some form of physiological incompatibility comparable with the different blood groups of man, and perhaps controlled by a genetic mechanism.

Drastic prenatal elimination on the scale occurring in the hare may well be more widespread among mammals than is generally realized. It certainly occurs in closely related species such as the rabbit and is even a feature of man, where it is likely that of all the zygotes formed, nearly half fail to develop to the point of eventual reproduction. Penrose estimates that this astonishing wastage is made up roughly as follows. Out of every 100 zygotes, about 3 are still births, about 15 die in the early prenatal period, 2 shortly after birth, and 2 at an early juvenile age, 16 remain unmarried, and 6 marry but do not reproduce. How much of this wastage can be attributed to genetical causes and how much to the environment is not known.

Natural selection and polymorphism

Sometimes the nature of a selective agent can be ascertained, as Cain and Sheppard have shown with conspicuous success in the brown-lipped snail, *Cepaea nemoralis.* The colour of the shell in this species varies greatly; it can be classified into five shades ranging from yellow through fawn and pink to brown. The markings on the shell also vary from unbanded forms at one extreme to those with five dark brown bands running round the shell at the other (Figure 25). Both characteristics are

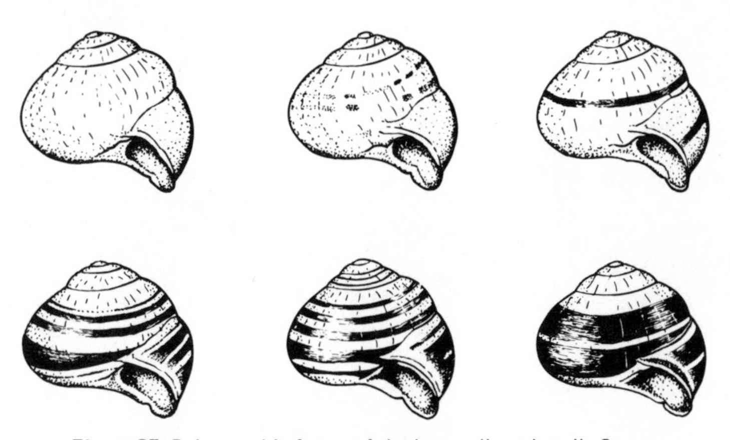

Figure 25. Polymorphic forms of the brown-lipped snail, *Cepaea nemoralis,* showing different degrees of banding

under genetic control and represent a balanced polymorphism (page 47). Sampling of snail populations has shown the proportion of yellow shells to be highest in those areas with the greenest background, such as open downland. Similarly it is lowest in the least green localities, and falls to zero in the thickest beech woods. The amount of banding also shows marked correlation with the uniformity of the surroundings. In the most uniform areas the proportion of unbanded shells is high, and in one region of downland beech wood reached 100 per cent. Among more variable surroundings banding becomes an advantage, and plain shells accounted for only 1.2 per cent of a sample taken from a hedge with a thick undergrowth of brambles and little green vegetation.

Cain and Sheppard have demonstrated the nature of a selective agent which plays a part in bringing about this striking degree of adaptation.

Figure 26. Thrush anvil stone with cracked shells of *Cepaea nemoralis* and *C. hortensis*

The chief predator on *C. nemoralis* is the song thrush, *Turdus ericetorum.* A large number of living snails were collected and each shell was marked with a dot of paint on the ventral side so that the colour would be invisible to birds. They were then liberated in a small copse. Subsequent sampling here and in another similar wood nearby showed a marked decrease in the proportion of yellow snails killed from the middle of April onwards as the background colour was becoming continually greener. Thus in mid-April yellow was at a disadvantage, the surroundings being relatively brown; by late April it was of neutral survival value, and by mid-May it was advantageous. Important independent evidence was provided by broken shells collected from thrush 'anvil stones' in the two woods (Figure 26). These showed the same fluctuation in the proportion of yellow individuals, which declined as the season progressed. Moreover, it was found that the rate of change was the same in the two localities and that selection by thrushes was thus similar in each.

The selection of a particular characteristic (or, rather, of the genes controlling it) must often involve a balance of advantage against disadvantage. Thus, as in *Cepaea nemoralis,* a character which is deleterious in one environment may well prove beneficial in another. Allison has studied a human blood condition known as sickle-cell anaemia which is restricted to parts of Africa, India, and the Mediterranean region, and is caused by an abnormality in the haemoglobin. The normal molecule (Hb^n) consists essentially of two groups of atoms, the 'haem' portion (known as the prosthetic group and including the iron atoms needed for oxygen transport) and the 'globin' portion (a sequence of amino-acids). In sickle-cell haemoglobin (Hb^a), one amino-acid in the 'globin' group (glutamic acid) is replaced by valine. Thus,

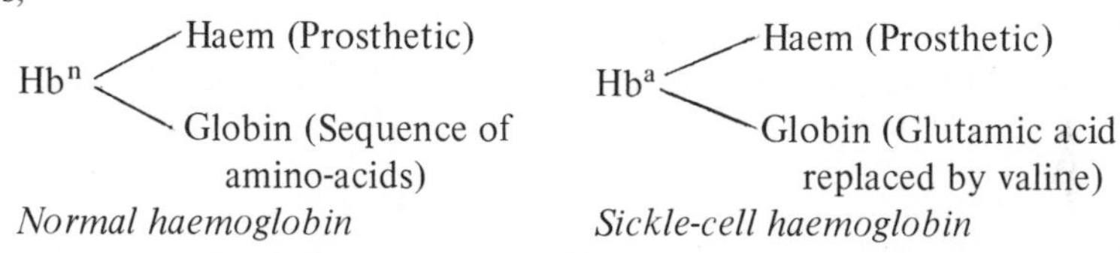

The change in the sickle-cell haemoglobin brings about an upset in electrical equilibrium, the molecules tending to aggregate when they become de-oxygenated. As a result the red cells collapse, assuming the characteristic sickle shape and losing their haemoglobin in the process (Figure 27). Anaemia follows, associated with other complications such

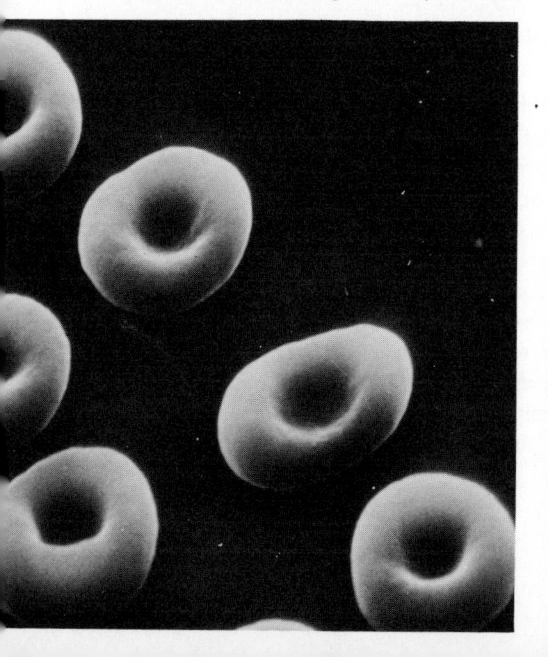
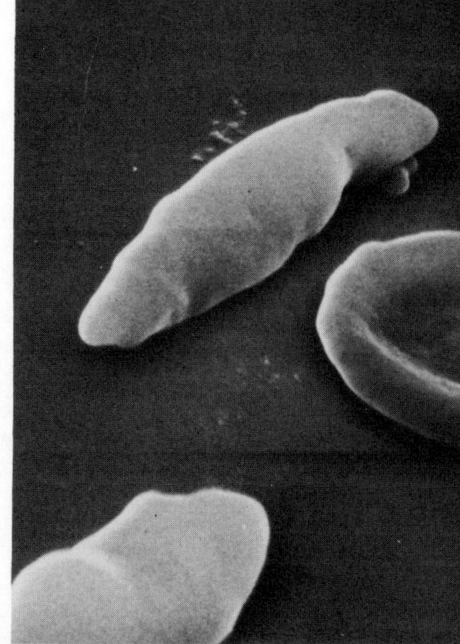

Figure 27. *(left)* Normal human red blood cells (x 4500) and *(right)* sickle cells (x 7000) showing the collapsed condition

as the blockage of small blood vessels and an upset in liver metabolism. The two forms of blood are controlled genetically by a pair of alleles, that producing the normal condition (*N*) being dominant to abnormal (*n*). Homozygous dominant individuals (*NN*) therefore carry normal haemoglobin (Hb^n) while homozygous recessives (*nn*) are anaemic, since they possess sickle-cell haemoglobin (Hb^a). Their expectation of survival to reproductive age is low, and it is estimated that they produce only about a quarter as many offspring as members of the other two groups. Heterozygous individuals (*Nn*) carry both kinds of haemoglobin (Hb^n and Hb^a), the behaviour of the normal type masking that of the abnormal. Such persons are said to exhibit the sickle-cell trait.

In spite of its disadvantages, sickle-cell anaemia is surprisingly common in certain parts of the world, and the homozygous recessive condition (*nn*) may reach a level of 4 per cent or more in some African populations. Using the Hardy-Weinberg formula (see Chapter 2) we can write,

Genotypes	*NN*	*Nn*	*nn*
Proportions	p^2	$2pq$	q^2
Percentages	64	32	4

The figures for percentages are derived as follows:

The incidence of homozygous recessive individuals (*nn*) = $q^2 = 0.04$ (4 per cent). Now $p + q = 1$ where p = the frequency of gene *N*, and q = the frequency of gene *n*, and so

$$q = \sqrt{0.04} = 0.2$$
$$p = 1 - 0.2 = 0.8$$
$$p^2 = (0.8)^2 = 0.64\ (64\%)$$
$$2pq = 2 \times 0.8 \times 0.2 = 0.32\ (32\%)$$

Thus, in order to maintain a balanced polymorphism in a population where 4 per cent of the children are born with sickle-cell anaemia (*nn*), some 32 per cent must be heterozygous for the gene (*Nn*) and 64 per cent normal (*NN*). In parts of Africa, it is estimated that the incidence of the sickle-cell trait attains a maximum level of 40 per cent, in India 30 per cent, and in Greece 17 per cent. Evidently, the disadvantages associated with Hb^a must be counterbalanced by some powerful selective advantages. Early investigations showed that areas with the highest incidence of sickle-cell were those with the worst record of malignant tertian malaria. In order to establish a relationship between the two it was necessary to sample appropriate populations. Adults are bad experimental subjects for this purpose owing to the immunity which builds up in their bodies after several attacks of malaria. Random samples of children were therefore selected as shown in Table 8.

Table 8. Relationship between sickle-cells in the blood and malaria parasites in children

	Malaria parasites in blood	No malaria parasites in blood	Total
Sickle-cells	12 (27.9%)	31 (72.1%)	43
No sickle-cells	113 (45.7%)	134 (53.3%)	247
Total	125	165	290

$\chi^2_{(1)} = 5.1$, for which $P \triangleq 0.02$. In other words there was a significant association between the incidence of the sickle-celled condition and lack of malaria parasites.

Two samples of adults with and without sickle cells in their blood were artificially infected with malaria parasites (*Plasmodium falciparum*) and their progress was followed for forty days afterwards (Table 9).

Table 9. Results of artificial infection with malaria parasites

	Developed malaria	No malaria	Total
Sickle-cells	2*	13	15
No sickle-cells	14	1	15
Total	16	14	30

* The density of parasites in these two cases remained very low in spite of repeated infection.

Despite the small numbers involved, the implications are clear enough.

The essential features of the sickle-cell polymorphism in man can thus be summarized as shown in Table 10.

Table 10. Features of the sickle-cell polymorphism in man

Genotypes	*NN*	*Nn*	*nn*
Phenotypes	Normal	Sickle-cell trait	Sickle-cell anaemia
Haemoglobin	Hb^n	Hb^n and Hb^a	Hb^a
Anaemia	Not anaemic	Not anaemic	Anaemic
Malaria	Not resistant	Resistant	Resistant

In areas with a high incidence of malaria the *nn* genotype will be at an advantage, whereas *NN* individuals will tend to be selected in non-malarious districts. Of particular significance are the heterozygotes which combine the advantages of both homozygotes without suffering the defects of either. Such *heterozygous advantage* (or *heterosis*) is a well known genetic phenomenon which can be of great importance in determining the adaptability of a species.

Individuals heterozygous for one pair of alleles are likely to be heterozygous for many others as well, so maintaining diversity within a population and promoting the evolution of the best adapted type. As we have seen above, fitness of the heterozygote is determined in relation to that of the two homozygotes, a situation which usually provides the basis of a balanced polymorphism.

Polymorphic situations such as that occurring in the primrose, *Primula vulgaris* (page 47), usually respond readily to changes in selection pressure, and we would expect this to be true of sickle-cell as well. Fortunately, we can put the hypothesis to the test, for the negro slaves transported from West Africa to North America some 300 years ago must have carried the sickle-cell trait at a level of at least 22 per cent. Today the incidence is 9 per cent, presumably due to the greater advantage of the *NN* phenotype in the absence of malaria. Allowing for a 7 per cent dilution due to intermarriage with whites and Indians, this leaves a 6 per cent reduction in *Nn* individuals in approximately 12 generations.

Natural selection and distribution

The situation in the brown-lipped snail, *Cepaea nemoralis,* already described illustrates an important general principle concerning the action of natural selection. Not only does its intensity fluctuate at different times within a single locality, but in a widely distributed species the effects may show a marked contrast in the various parts of its range.

Studies of spotting on the hind-wings of the meadow brown butterfly, *Maniola jurtina,* from 1950 onwards, have revealed a most interesting situation in which the females from southern England exhibit two distinct forms (Figure 28).

Those extending eastwards from Devon to Kent have a spot-distribution unimodal at 0 (Southern English stabilization), while samples from east Cornwall are bimodal with the larger number at 0 and the smaller at 2 spots (East Cornish stabilization). It might be expected

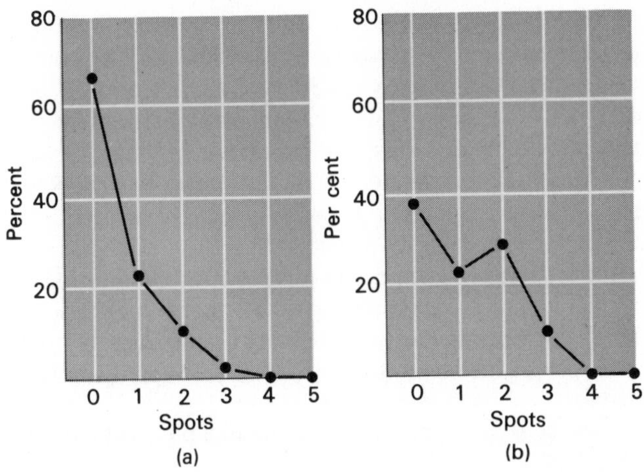

Figure 28. Spot-distribution in females of the butterfly, *Maniola jurtina:* (a) Southern English and (b) East Cornish

that the change-over from one form to the other would be a gradual one—in other words, a cline, but this is not so. The characteristic features of the transition zone are as follows:

(i) The change from unimodality to bimodality in the females may take place with surprising abruptness, sometimes within only a few yards.
(ii) The location of the boundary between the two forms has fluctuated somewhat since it was discovered in 1956, moving westwards and then eastwards again over a distance of approximately forty miles during a period of thirteen years.
(iii) No physical barrier is involved in separating the Southern English and East Cornish populations.
(iv) The situation in the central area of the Cornish peninsula just described, is repeated further south and is therefore not peculiar to a single locality.
(v) The essential features of the situation have remained unchanged (within the limits mentioned in (ii)) for fifteen years and probably much longer.

The aspects outlined above serve to highlight two important generalities. First, there has evidently been a remarkable stability in spotting, a situation already mentioned in Chapter 3 (page 56). We now know that this represents not just an isolated Southern English stabilization, but the western extremity of a General European one extending south-eastwards for some 3000 miles. In Cornwall, the insect is approaching the western extremity of its range (it is absent from North America), and an increased sensitivity to ecological changes could well account for the diversity of adjustments that it makes to different environments. As we shall see, this diversity achieves maximum expression in the Isles of Scilly (Chapter 5).

Secondly, the transition from one form of spot-distribution to another in the absence of a physical barrier, indicates that powerful selective forces must be at work in the two adjoining populations, restricting to a considerable extent the amount of gene-flow between them. Groups of organisms separated from one another in this way are said to exhibit *sympatry.* On the other hand, isolation due to a physical barrier results in *allopatry.*

In attempting to interpret the situation in the boundary area, it is clear that we are faced with the operation of selective forces of a magnitude and nature unlike anything we have encountered so far. Observations in the field have contributed relatively little, beyond the knowledge that mating between individuals with different spot-values appears to be at random, and that predation on the adults is not a major factor in selection.

For a closer investigation it was necessary to study the animal under laboratory conditions, a first requirement being to discover the principal causes of mortality. The greater part of the life cycle of the insect is spent as a larva (Figure 29)–roughly from August of one year to June of the next. It therefore seemed reasonable to assume that an appreciable part of selective elimination might take place during this phase. Implicit in this assumption was the idea that some factor affecting larval survival might be related to the spot-values of the resulting adults. Unfortunately, overwintering involves hibernation at the second or third larval instar, and it has proved difficult to collect wild larvae until they resume feeding about the beginning of May. They can then be swept with a net from grass stems at night and reared in breeding cages in the laboratory. From these experiments we now know that larval mortality is due to two main causes.

The more obvious of these is infestation by the Braconid, *Apanteles tetricus* (Figure 30), a minute Hymenopteran, similar in appearance to *A. glomeratus* which parasitizes and controls the large white butterfly, *Pieris brassicae.* Eggs are laid inside the body of the larva probably during

Figure 29. Larva of the meadow brown butterfly, *Maniola jurtina* (x 1½). The body is green with a thin yellow line along each side

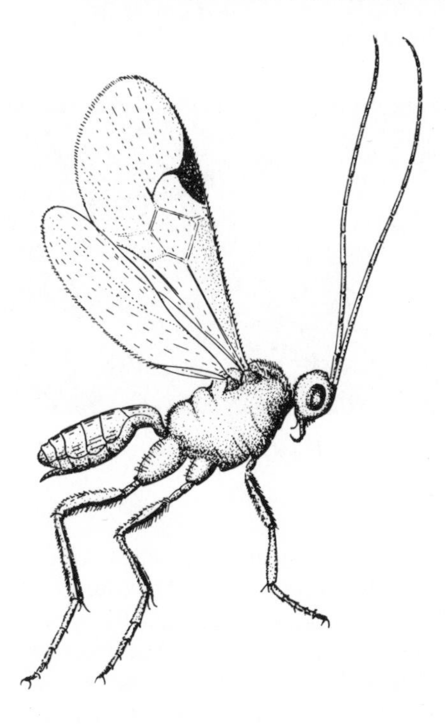

Figure 30. The Braconid, *Apanteles tetricus,* a parasite of the larva of the meadow brown butterfly, *Maniola jurtina* (x 25)

the first instar, and the mature grubs eventually burrow out of the haemocoel and through the body wall of the host when it is fully grown and about to pupate. Evidence of *Apanteles* parasitism is thus easy to detect, for the dying host is surrounded by a number of small white cocoons (Figure 31). Deaths due to *Apanteles* among larvae collected during May seldom attain a level higher than 5 per cent and are generally far fewer. Those occurring in larvae obtained by sampling during June and July may amount to 75 per cent or more.

A second and more elusive cause of larval mortality is infection by bacteria. Among larvae collected in May, this can account for 95 per cent or more of the deaths, but only about 25 per cent of the caterpillars swept in June and July die of this cause. The bacterial syndrome is easy to identify, the body of an affected larva blackening from the hind end forwards and death occurring in two or three days. The pathology is more complex than was at one time supposed and still remains largely unsolved. Suffice it to say that all infections are associated with one or more species of *Pseudomonas,* but whether this is the primary pathogen or merely a later colonist of an already infected body is not certain.

Figure 31. Fully grown larva of *Maniola jurtina,* surrounded by cocoons of the Braconid, *Apanteles tetricus* (x 2)

Comparison of spot-distributions in reared and flying females revealed some interesting differences between them (Figure 32). The adults that emerged from late June onwards exhibited a characteristic pattern of spotting. Among the males, the spot-distribution was the same as that of their wild counterparts–unimodal at 2 spots. Of the females, those that emerged early (roughly from mid-June–mid-July) showed an excess

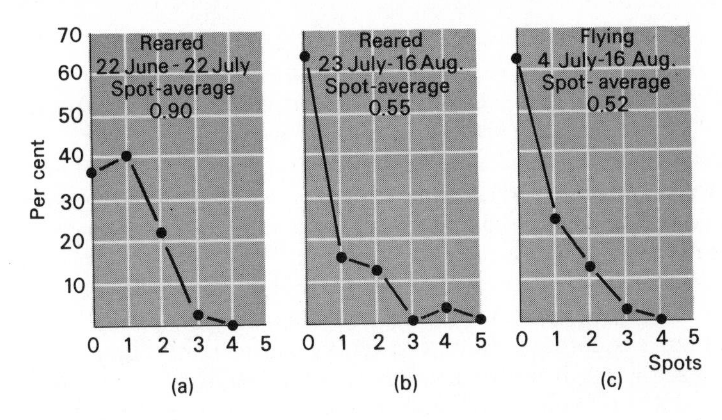

Figure 32. Comparison of spot-distribution in reared and flying females of the butterfly, *Maniola jurtina,* from a colony near Andover (Hampshire) in 1959

of spots (i.e. a high spot-average, see Figure 32), but as the emergence proceeded spotting declined, tending to resemble that of wild insects. Figure 32 shows a typical set of data derived from a colony near Andover where much of this work was done. Comparison of the spot-distribution in females emerging early (a) with those flying in the same locality (c) gives $\chi^2_{(2)} = 9.76$ ($P = 0.01$ to 0.001). However, a similar comparison of those emerging later (b) shows $\chi^2_{(2)} = 3.08$ ($P = 0.3$ to 0.2). On the basis of these results, it might be expected that in wild colonies, the earliest females to appear would be more highly spotted than those emerging later. Such a situation has been detected on a number of occasions and may well be universal. That it has not been encountered more commonly is probably due to the fact that in most populations the transition from high spotting to low takes place rather rapidly while the insect is still scarce, and it is therefore difficult to procure large enough samples for statistical analysis.

Laboratory studies of *Maniola* thus throw some light on the significance of adult spotting, indicating strongly the endocylic nature of selection, in that it operates differently in various phases of the life cycle (see also page 65 for evidence of a similar situation in the hare and in man). So far, little attention has been paid to the study of the egg and pupa stages beyond establishing that an appreciable mortality occurs in both. But whether elimination is selective (as it probably is) and if so, whether it affects adult spot-distribution, is still not known.

From the results of breeding, we see that selection for spotting in adult, female meadow browns operates in two stages—an early anti-low-spot phase and a later anti-high-spot one. The first of these corresponds to the period of high bacterial mortality in the larvae and the second to greater elimination by *Apanteles.* It was tempting, therefore, to explain spot-stabilization in the boundary area and elsewhere in terms of a balance of selective forces, bacteria having the effect of increasing spot-values and *Apanteles* tending to reduce them. Such a hypothesis is, fortunately, easy to test by rearing larvae obtained from adult pairings made in the laboratory, out of contact with *Apanteles.* The resulting emergence of adults followed the pattern obtained previously. Evidently, the Braconid is not a primary selective agent where spotting is concerned, and the fact that larvae swept late exhibit a high infestation can probably be attributed to the effect of the parasite in slowing down their rate of growth. Evidence for the influence of bacteria is more convincing, for when stocks of larvae are reared from known pairings and accompanied by an exceptionally high incidence of bacterial infestation, the level of spotting in the resulting adults has been found to exceed the value that might be expected from the parents concerned. However, much further study is needed, particularly on the nature of bacterial ecology in grassland, before the mechanism of spot-selection can be fully worked out. In conclusion, it should be noted that most of what has been said so far refers to the females of *Maniola jurtina.* The fact that in the males there is little distinction in spotting between reared and wild adults is of some interest, for it suggests that the adaptive significance and method of control of spotting may well be different in the two sexes.

Calculation of the selection pressure operating in nature against highly spotted female *jurtina* from the Andover colony, shows that selective elimination of individuals with 2 spots and over compared with unspotted adults was of the order of 70 per cent. In other words, a larva or pupa whose genotype would cause it to become a female with 2 spots or more had approximately a 70 per cent chance of being eliminated compared with one destined to develop as a female with 0 spots. The existence of such high selection pressures might well account for the abrupt transition in spotting on the borders of Devon and Cornwall. It also raises an important genetic question. If the selective disadvantage of high female spotting is so great, why have not the gene systems concerned disappeared altogether? A plausible explanation may be that while the genes for spotting are disadvantageous in the female, they are advantageous in the male (particularly the two-spot condition). Thus, as fast as the spot-genotypes are eliminated in the female, they are re-introduced for the next generation by the male in pairing. Alternatively, the genes for spotting may have advantages balancing the effect of bacterial disease. Support for such a hypothesis is provided on the one hand by the situation in Ireland where the genes for female spotting have been almost eliminated, and by that in East Cornwall on the other where high-spot values are at a relative advantage.

Judged superficially, such a system appears to be extraordinarily 'inefficient', but it is hardly more so than the many polymorphisms such as human sickle-cell (page 69) which maintain in the heterozygous state genes that can be disadvantageous when homozygous.

Selection and clines

Among a widespread and variable species scattered within a diverse environment we might expect to find a continuous gradation of characteristics in response to the changing conditions throughout its range. Such gradients of variation within distinct zones occur widely among animals and plants and are known as *clines.* Sometimes these take place over a relatively short distance and in response to a distinct series of varying ecological conditions (such clines are often called *ecoclines*). For instance, Gregor showed that the length of the flower stem of the sea plantain, *Plantago maritima,* a common halophytic (salt tolerant) plant, varies with the nature of the soil in which it lives. Inhabitants of waterlogged coastal mud had an average scape-length of just over 20 cm, while in those living on cliff faces the scape-length averaged nearly 50 cm. When cultivated in the same garden soil both types were found to retain their former characteristics, thus indicating that these were not merely localized environmental effects but inherent potentialities established by the action of natural selection.

Many of the clines which have been studied in detail occur on a much larger scale, such as that of the Atlantic fulmar, *Fulmarus g. glacialis* (Figure 33). J. Fisher found that the birds could be classified into four groups on the basis of coloration, ranging from the lightest forms with head, neck, and underparts all white, through two grades of intermediates, to the darkest individuals, whose whole body is of a

Figure 33. The Atlantic fulmar, *Fulmarus g. glacialis* (light form)

uniformly deep blue-grey colour. The proportion of these dark birds in the different populations showed a marked decline from north to south, ranging from 95 per cent in north-east Greenland to about 50 per cent further south. Off Iceland the percentage fell further to a maximum of 20; in Britain it was 2 or less (Figure 34).

In many warm-blooded animals it is well known that the intensity of melanin pigmentation tends to decrease with mean temperature, but to increase with relative humidity (Gloger's rule). Attempts to correlate coloration in the fulmar with humidity have, however, been unsuccessful, for, as Fisher pointed out, the zone of 85 per cent humidity or more includes the typically light-coloured populations of St. Kilda, Shetland and the Faeroes. Salomonsen suggested that there might be some connection between colour and the temperature of the sea water. Thus the dark forms of the Atlantic fulmar are practically confined to those areas where the surface temperature of the sea is at or

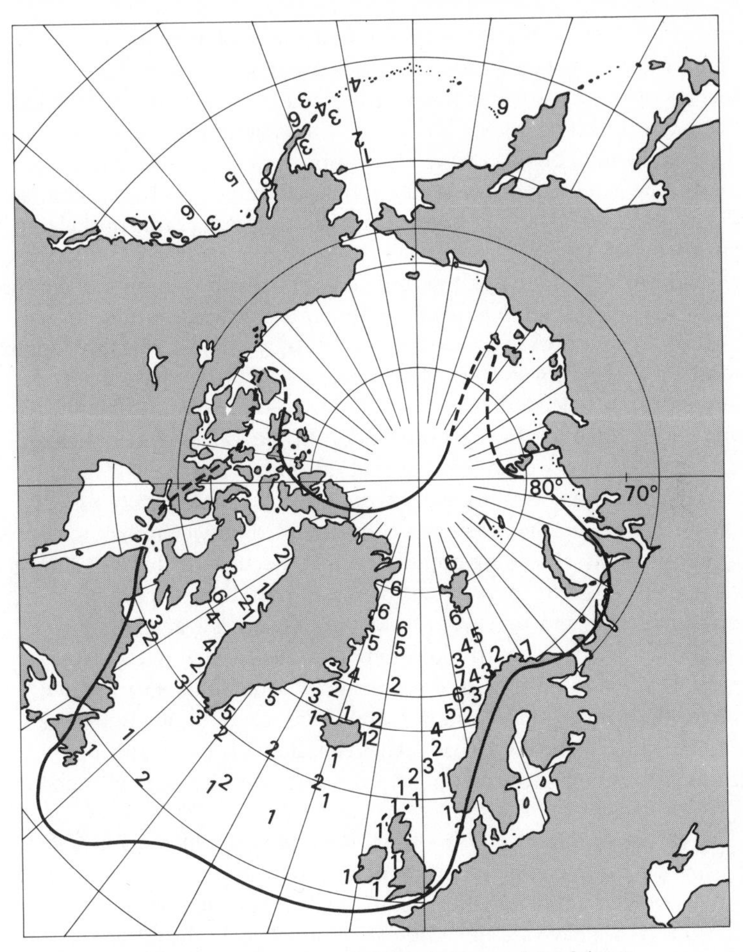

Figure 34. Distribution of the dark forms of the fulmar, *Fulmarus g. glacialis,* at sea, May to August.

Key:
1. 0-2%
2. 2-20%
3. 20-40%
4. 40-60%
5. 60-80%
6. 80-98%
7. 98-100%

near freezing point even in July and August ('high Arctic'). On the southern borders of this region the populations are mixed, while in the 'low Arctic' and Atlantic, where the water is warmer, only light individuals are found. The proportions of the various forms in the different regions are so characteristic that they would seem to be stabilized in some way by the process of natural selection. The relationship between water temperature and pigmentation is convincing, and this may well be one of the selective associations involved. But as yet we know nothing of two of the most important aspects of this cline, namely, the genetic mechanism by which variation in colour is controlled and the selective advantage enjoyed by the different forms in their respective zones.

Situations of the kind just described are probably widespread among populations of animals and plants with a more or less continuous distribution over a wide area, and, indeed, they are just what we would expect on grounds of evolutionary theory. As yet they have been comparatively little studied, no doubt due to the enormous labour involved in collecting the necessary data. In the longer clines, such as that of the fulmar, we can picture a situation where the environments at the extremities may differ greatly from one another and may reflect their differences in the organisms concerned. In between, a more or less continuously interbreeding population may occur, but this will almost always be isolated to some extent owing to partial barriers such as mountain ranges and forests, or unfavourable zones where the population density is greatly lowered. Thus, as Huxley put it, clines represent a 'partial biological discontinuity'. Their existence frequently leads to the formation of varieties, and more rarely to new subspecies between which some degree of sexual isolation has been built up; but varying degrees of inter-breeding ensure that even the most extreme forms will never attain the status of true species.

One of the problems that has beset the study of clines is a lack of information on the mode of action of selective agents, for it is often far from clear what characteristics are at an advantage in a particular situation. As we have seen in *Maniola,* spot values in the adult are probably indicative of advantages enjoyed by the larva and play little or no part in the life of the mature insect. A similar situation could well obtain in the fulmar, where the various colour forms may be associated with other advantages and disadvantages, possibly of a physiological kind. Earlier in the chapter we saw that selective elimination can attain a value of 70 per cent—far higher than was previously supposed. Pressures of this magnitude operating in different directions at the extremities of a cline could well account for the occurrence of sympatric evolution.

Most of the clines studied so far have extended over relatively long distances and embraced a wide variety of environmental conditions. A precise analysis of the variation occurring within them has, therefore, been impossible. An intensive study centred on a small, isolated population could produce useful information, both on the mode of action of selective agents and on the establishment of sympatry in the face of appreciable gene-flow. The outcome of such studies will be considered in the next chapter.

5

Natural Selection and Gene Flow

Selection and isolation

One of the most potent factors promoting evolution and the production of new species is isolation. This may assume a number of forms, depending on the kind of ecological barriers involved and hence the extent to which they serve to restrict gene-flow. For instance, isolation may be *geographical* as when land masses become subdivided by tracts of sea, mountain ranges or great rivers, to mention only a few possibilities. A preliminary to true speciation will be the formation of distinct geographical races or subspecies, and many examples of these are known. Among plants the yellow gentian, *Gentiana lutea,* whose extract is used medicinally as a tonic, is widely distributed in central Europe as the subspecies *G. l. lutea,* which has free anthers. In the Balkans, however, the form *G. l. symphandra* occurs exclusively, being mainly distinguishable by its united anthers. A slight amount of overlap exists between the two in a short intermediate zone but otherwise they are distinct.

In Britain an instance of geographical isolation among birds is provided by the wren, *Troglodytes troglodytes,* which forms distinct subspecies on St. Kilda, in the Shetlands, and on other Outer Hebridean islands. Their characteristics are summarized in Table 11.

The three subspecies show marked divergences in colour and size, both from one another and from the mainland stock. The variation in wing-length is particularly striking when expressed graphically (Figure 35). It will be seen that the difference between the extremes from St. Kilda and the mainland is so great that they do not even overlap. As with so many similar instances of adaptation, we know nothing as yet of the evolutionary significance of localized variations in size and colour, nor, indeed, whether they are associated with any corresponding physiological advantages. Suffice it to say here that the

Table 11. Geographical variation in three subspecies of the wren, *Troglodytes troglodytes*

Subspecies	Locality	Wing-length	Appearance
T. t. hirtensis	St. Kilda	51-55 mm	Greyish-brown above, paler under-parts.
T. t. zetlandicus	Shetland Is.	50-54 mm	Darker and more bulky in appearance than *T. t. hirtensis.*
T. t. hebridensis	Outer Hebrides	48-53 mm	Similar to *T. t. zetlandicus* but under-parts more buff-coloured and barring less heavy.

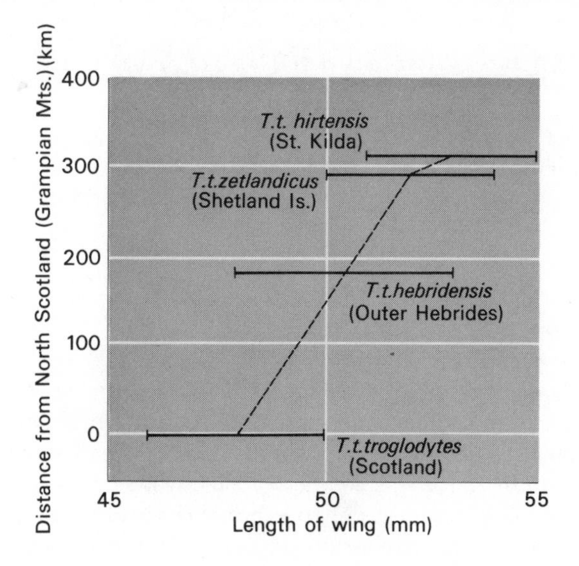

Figure 35. Isolation and size variation in the wren, *Troglodytes troglodytes.* (After Salomonsen.)

situation in the wren serves to illustrate a biological principle which applies to many warm-blooded animals, namely, that within a variable species the body size of its subspecies usually increases with decreasing mean temperature of their habitats (Bergmann's rule). For the wrens, it has been shown that a 1 per cent change in wing-length corresponds to little more than 0.5 °N latitude, while for the redpoll, *Carduelis*

Figure 36. Common marbled carpet moth, *Dysstroma truncata* (x 2). Pairing with the dark marbled carpet, *D. citrata,* and the production of fertile offspring can occur in the laboratory but not under natural conditions

flammea, a difference of 2° is needed to achieve the same effect, and in puffins, *Fratercula artica,* just over 1°.

Sometimes isolation is of a *seasonal* kind, as when two forms occupying the same or kindred ecological niches are capable of interbreeding but in fact never do so on account of the different times of their appearance. Thus, among moths in Britain the common marbled carpet, *Dysstroma truncata,* is double brooded, while the closely related marbled carpet, *D. citrata,* has only one brood (Figure 36). In nature the two seldom if ever pair, as the adults of *citrata* are on the wing between the two broods of *truncata.* Under experimental conditions the two can be brought together by artificially slowing down the development of one or accelerating that of the other. Pairing then takes place readily with the production of fertile offspring. Evidently the two species are closely related and may well have evolved from one, but how this situation has come about is not known. It could perhaps have resulted from geographical isolation at some time in the remote past.

Again, isolation may be *ecological* and result from the peculiar environmental conditions required by a species for its successful survival. Sumner made a detailed study in Florida of the three subspecies of the mouse *Peromyscus polionotus* which occur there. *P. p. leucocephalus* is an island form, almost pure white in colour and matching its background of sand. Its appearance has evidently been stabilized by selection, and it shows no tendency to vary throughout its range. *P. p. polionotus* is a dark inland subspecies, also invariable and blending with the darker surroundings. The third form, *P. p. albifrons*, also inhabiting the mainland, is intermediate in colour between the other two and much more variable. The situation is summarized in Table 12.

Table 12. Variation and distribution in the mouse, *Peromyscus*

Subspecies	Colour	Variation	Habitat
P. p. leucocephalus	Nearly white	Invariable	Coastal. White sand and sparse vegetation.
P. p. albifrons	Pale to darker brown	Variable	Coastal. White sand and sparse vegetation extending inland to darker soil with increased vegetation.
P. p. polionotus	Dark brown	Invariable	Inland. Dark soil and extensive vegetation.

The variable appearance of *P. p. albifrons* can be related to the nature of its habitat which is subject to considerable changes. Thus, on the shore the conditions are those characteristic of *P. p. leucocephalus*, while inland, the darkening background favours a corresponding change in colour. The region of overlap between *albifrons* and *polionotus* about forty miles inland is quite narrow (only a few miles), and thereafter only the dark subspecies is to be found. A Lamarckian interpretation of the situation is ruled out by the results of laboratory breeding experiments which showed that:

(i) the three forms maintain their identity when reared under similar environmental conditions,

(ii) segregation for intensity of pigment occurs in F_2 families, indicating its control by quite a simple genetic mechanism.

The situation in *Peromyscus* thus accords well with our ideas of natural selection, for we would expect just such a rapid change-over from one subspecies to another in an area where neither is well adapted to match its surroundings, and both are therefore subject to the maximum elimination by predators.

In certain circumstances evolution may result from the combined action of ecological and geographical isolation working together, as Camin and Ehrlich have shown for the water snake, *Natrix sipedon*, inhabiting the Lake Erie islands. The species is characterized by a considerable degree of colour variation (Figure 37) which can be divided

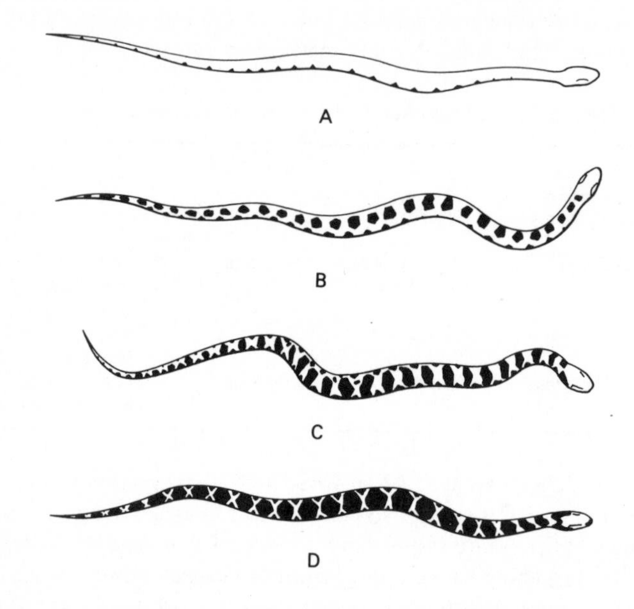

Figure 37. Diagrammatic representation of the colour variation in the water snake, *Natrix sipedon*

for convenience into four categories ranging from unbanded, light forms (A) to dark, banded individuals (D). Although a certain amount of migration by swimming undoubtedly occurs from island to island, also from the Canadian mainland in the north (Ontario) and from Ohio (U.S.A.) in the south, the various populations are nonetheless relatively isolated from one another (Figure 38).

The results of sampling the water snake colonies on the mainland and on the different islands are summarized in Figure 39. From the histograms it is clear that while banded forms (type D) predominate on the mainland to the north and south, their incidence on the islands is

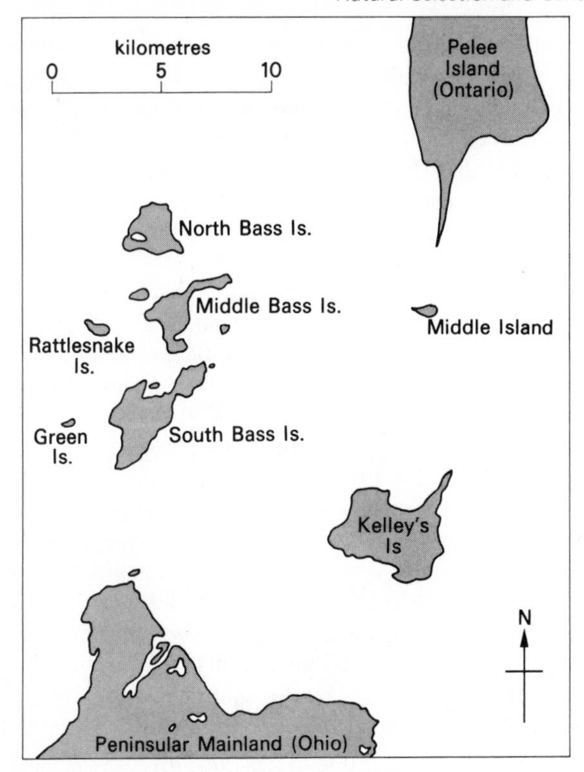

Figure 38. The islands of Lake Erie and their relationship with the mainland of the U.S.A.

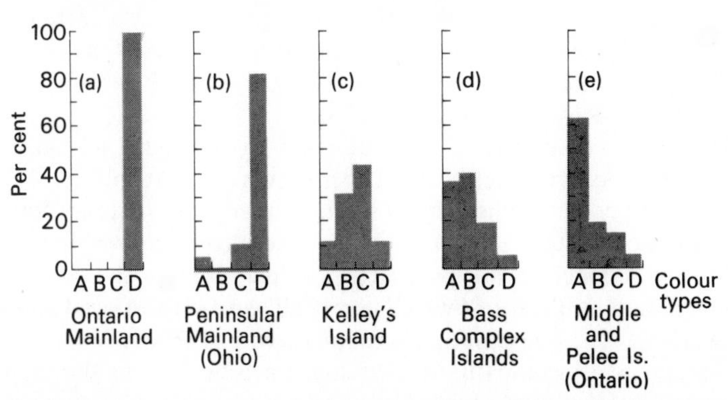

Figure 39. Colour variation in water snakes of Lake Erie. (After Camin and Ehrlich.)

much reduced in favour of the other three kinds. The possibility that these colour differences might merely result from environmental effects is disposed of by breeding experiments, which show that the litters of unbanded parents contain a high proportion of unbanded young. It is equally impossible that the abundance of unbanded individuals might be maintained by migration from elsewhere, for the only nearby populalations which could supply migrants are on the mainland and these are typically dark coloured.

Genetic drift (see page 99) might be a significant factor here, but this can be excluded for two reasons. First, the snake populations are relatively large (seven collectors captured 400 *N. sipedon* on Middle Island in five hours). Second, the trend on every island is in the same direction–towards a reduction in banding. This strongly suggests a systematic selection pressure rather than random gene survival.

Camin and Ehrlich point out that to the human eye unbanded snakes are particularly well concealed among the flat, limestone rocks of the islands. By contrast, banded individuals are highly conspicuous in these conditions. Moreover, there is good evidence of visual predation by birds such as gulls and herons which are common in the area. The total elimination of the genes causing banding is, however, prevented by a continual influx of dark mainland immigrants. We thus have the interesting situation of natural selection and migration working in opposition, the former tending to reduce banding, the latter to increase it.

Sometimes isolation is *sexual,* natural selection having favoured the evolution of physiological and structural barriers to cross-fertilization in the process of species formation. Here belong the many instances of interspecific incompatibility among plants, both as a result of physiological failure in the process of fertilization and on account of mechanical devices preventing pollination.

Demerec has described the crossing of different varieties of maize, *Zea mais.* If the popcorn variety is used as the female parent and non-popcorn pollen applied to the stigmas, hardly any seeds are formed. Pollination of the type *popcorn* × *popcorn* resulted in normal seed production. Popcorn plants pollinated with a mixture of the two kinds of pollen produced seeds of which nearly all were self-fertilized and very few were hybrids. The situation is summarized in Table 13.

This apparent incompatibility was confirmed by dividing the stigmas of an ear of popcorn into two groups and pollinating each with one kind of pollen. The expected result was then obtained, namely, that the cross

Table 13. Results of three crosses in maize. (After Demerec.)

Cross: Female (stigma)	Cross: Male (pollen)	Seeds formed
popcorn	*popcorn*	*popcorn*
popcorn	*popcorn* } mixture	*popcorn* (predominantly)
	non-popcorn } mixture	hybrid (very few)
popcorn	*non-popcorn*	hybrid (very few)

popcorn x *popcorn* gave a full complement of seeds, while that of *popcorn* x *non-popcorn* produced hardly any. The reason for the failure of the 'foreign' pollen is not certain, but in many plants, including some varieties of maize, it is known that the growth rate of the pollen tubes down the style in abnormal pollination is slower than in normal, and may fail altogether. Thus if two kinds of pollen reach a stigma together, the normal form will always be at an advantage and more likely to achieve fetilization. A somewhat comparable situation has already been described in the heterostyly of the primrose, *Primula vulgaris* (page 47).

Among animals sexual isolation of species is often characterized by the development of specific scents, colours, sounds, and behaviour patterns, and also the possession of peculiar copulating devices such as are found in many insects with elaborate external genitalia.

Darwin held the view that within a single species all bright colours or other display devices, such as song, specific to one sex or the other (generally the male) must be regarded as conferring an advantage on their possessors in securing a mate. The emphasis originally laid on this special aspect of natural selection, sometimes referred to as *sexual selection,* is now known to have been exaggerated. Field observations have established beyond doubt that many types of coloration and display in fact have other significance, while song in birds serves the primary function of threatening other intruding males rather than of attracting females. Similarly, the red breast of the robin, whose sexual significance had generally been assumed, has been shown by Lack to be solely concerned with threat display towards territorial rivals.

Tinbergen has demonstrated a comparable situation in fishes such as the three-spined stickleback, *Gasterosteus aculeatus.* With the onset of the breeding season the males assume nuptial colours, the eye becoming blue, the back changing from dull to greenish brown and the underparts developing a brilliant red colour. The changes in appearance are associated with a characteristic threat behaviour towards other males, which is the signal for an attack if the intruder fails to withdraw. Under experimental conditions in an aquarium, the same hostility is shown towards a model provided a red underside is visible. The size and shape of the model appear to be unimportant, and Tinbergen records an occasion when mature male sticklebacks even 'attacked' a red mail van passing the laboratory window about a hundred yards away. As the van went by the row of twenty aquaria, all the males dashed towards the side nearest the window and followed the van from one corner of their tank to the other.

Nonetheless, many instances remain where the function of colour, usually in association with characteristic behaviour patterns, is unquestionably that of stimulating the other sex to pairing and copulation. Such coloration is known as *epigamic* and is typified by the plumage of the male argus pheasant and peacock. Many of our own birds adopt a characteristic breeding plumage which is brighter than that assumed at other times of the year and of significance in courtship. Typical examples are the male mallard and bullfinch, whose respective females are relatively sombre in appearance (*cryptic colours*) thus matching their surroundings when sitting on the nest. Newts and many insects also provide familiar examples of a similar sexual dimorphism.

Sexual selection is undoubtedly widespread among vertebrates and in many invertebrates as well. In mammals it mainly takes the form of increased pugnacity among males during the breeding season and fighting for the possession of the females. The selective process involved would clearly account for the evolution of special secondary sexual characteristics, such as the antlers of the male deer. It is more difficult to understand how the individual preferences of females could bring about the development of such male characteristics which must be of little or no survival value for the greater part of the year.

Here is obviously a large field for further research, and in the absence of factual evidence, we can only conclude that the advantage derived by one half of a species from sexual selection at a vital period in its life history is of such magnitude as to outweigh any corresponding disadvantages which may accrue. Such situations must frequently arise in nature in which the operation of an advantageous character involves a balance of advantage and disadvantage (see also Chapter 4, page 70).

The occurrence of preferential mating has been demonstrated experimentally in many animal species. Sheppard investigated the behaviour of the scarlet tiger moth, *Panaxia dominula,* in which the effect of a single gene showing no dominance is recognizable in the heterozygote as the variety *medionigra,* and in the rarer homozygote as

Table 14. Preferential pairing in the scarlet tiger moth, *Panaxia dominula*

	Genotypes of the three moths in the cage	Result of first mating: Like genotypes pairing	Result of first mating: Unlike genotypes pairing	Totals
1	*dominula* ♂ *dominula* ♀ *medionigra* ♀	8	20	28
2	*medionigra* ♂ *dominula* ♀ *medionigra* ♀	12	14	26
3	*dominula* ♂ *medionigra* ♂ *medionigra* ♀	13	14	27
4	*dominula* ♂ *medionigra* ♂ *dominula* ♀	11	22	33
5	*medionigra* ♂ *medionigra* ♀ *bimacula* ♀	2	0	2
6	*bimacula* ♂ *medionigra* ♀ *bimacula* ♀	0	1	1
7	*medionigra* ♂ *bimacula* ♂ *bimacula* ♀	3	15	18
8	*medionigra* ♂ *bimacula* ♂ *medionigra* ♀	2	10	12
9	*dominula* ♂ *dominula* ♀ *bimacula* ♀	2	1	3
	Totals	53	97	150

bimacula. His results are summarized in Table 14, from which it is clear that unlike genotypes tend to pair more readily than similar ones. Observations made under laboratory conditions suggest that mating is controlled by the female, for on several occasions she was seen to reject a male of her own kind and subsequently accept one of a different form. In the absence of competitors, a male of any sort would usually succeed in eventually pairing, but courtship was longer between like than unlike genotypes.

Although such disassortive mating in *P. dominula* would be difficult to observe in nature, and has not often been recorded, it seems likely that it must occur to some extent. For, as Sheppard points out, even if a female of any genotype is equally attractive to all males, the control she exercises is bound to have some effect, as she often assembles a large number of males around her before pairing takes place.

Between many animal species a system of incompatibility may exist analogous to that already described in plants. It is well known that sperms, particularly of the higher animals, are extremely sensitive to changes in their environment, notably those of osmotic pressure, and in adverse circumstances their length of life is much reduced. This is no doubt one reason for the breakdown of interspecific crosses, namely, the death of the sperms before they reach the eggs.

On occasions where interspecific fertilization is successfully achieved the resulting offspring are generally inviable or at some disadvantage when compared with either parent. In flax plants, *Linum* spp, the seeds resulting from the cross *L. perenne* (female) x *L. austriacum* (male) are only able to germinate if the embryo is artificially freed from the seed coat. Left on their own, the young plants are incapable of breaking out of the testa and so die. The effects of the cross *L. austriacum* (female) x *L. perenne* (male) are even more extreme, for not only are the embryos incapable of freeing themselves, but they also require artificial germination in dilute sugar solution during the early stages if they are to survive. Many similar examples are known among animals, all of which must tend to enhance genetic isolation in the wild state. Thus the cross between the large and small elephant hawk moths, *Deilephila elpenor* (female) x *D. porcellus* (male) gives only male hybrids, the females, which are present as larvae, dying in the pupal stage. The reciprocal cross, however, results in sterile hybrids of both sexes. This illustrates an important general principle (sometimes known as Haldane's Law), namely that if, in a cross between two distinct species or races, one sex is absent, rare, or sterile, it is always the heterogametic sex (i.e. the female in Lepidoptera).

Isolation may also be achieved as a result of *genetic* causes. In such circumstances its attainment is generally, but not always, relatively rapid. The occurrence in the wild state of chromosome deficiencies (deletions) and duplications involving the loss or multiplication of genes, has been studied extensively by Dobzhansky in various populations of *Drosophila.* Such abnormalities, as might be expected, are generally accompanied by detectable phenotypic effects, unlike the results of inversions and translocations (page 32), which are not necessarily different from the parental form, since the total number of genes is unchanged, and only their relative positions on the chromosomes differ. For example, the shape of the Y chromosome in *D. pseudo-obscura* is subject to marked changes. In some strains it is large and V-shaped, in others one or both arms are shortened. Sometimes it is J-shaped and much reduced in length. In all some seven distinct forms have been detected and the distribution of their possessors has been mapped. Three of these have succeeded in colonizing a wide area but the remaining four are all restricted to distinct localities. Thus one strain occurs only in southern California, another is restricted to the highest part of the Rocky Mountains in northern Colorado, while a third is found only around Puget Sound, in the north-west. As Dobzhansky points out, it seems likely that all seven types have been derived from a common ancestor and, in some instances at least, the gene changes resulting from chromosome abnormalities have been responsible for influencing the ecological requirements of the species. Similar variations in the Y chromosome have also been detected in *Drosophila ananassae* and *D. simulans*.

Isolation and population size

One of the most obvious results of isolation is to limit the size of a population and hence the variety of genes which it can possess. Clearly, the greater the numbers the better will be the chance of different genetic factors being incorporated in new genotypes with the possibility of beneficial variation. Periodic fluctuations in numbers, well known among animal communities and some plants as well, must also be advantageous for this reason. On theoretical grounds we are thus led to the conclusion that the *potentiality* for evolution is greater in large populations than in small ones because:

(i) they afford better opportunities for the spread of genes having small but beneficial effects,
(ii) they are able to hold more genes in reserve whose influence in existing circumstances is neither beneficial nor harmful, but which might prove of adaptive value under changed environmental conditions,
(iii) the possibility of beneficial mutations occurring is increased.

Such differences in the rate of evolution would be difficult to demonstrate experimentally for many reasons, not the least being the problem of controlling environmental conditions with sufficient precision. So far investigations have been confined to observations of fluctuating populations of species such as the marsh fritillary butterfly, *Euphydryas aurinia,* in which it has been shown that outbursts of variation occur when the numbers are increasing. Furthermore, when the density decreases once more, the resulting form may be appreciably different in appearance from that which originally existed.

This is not to suggest that large populations of a species are more variable than small ones. On the contrary the reverse is usually true, for large areas will generally provide a greater variety of edaphic and climatic conditions than smaller ones. Thus big communities will tend to evolve a somewhat generalized form adapted to the average of the conditions

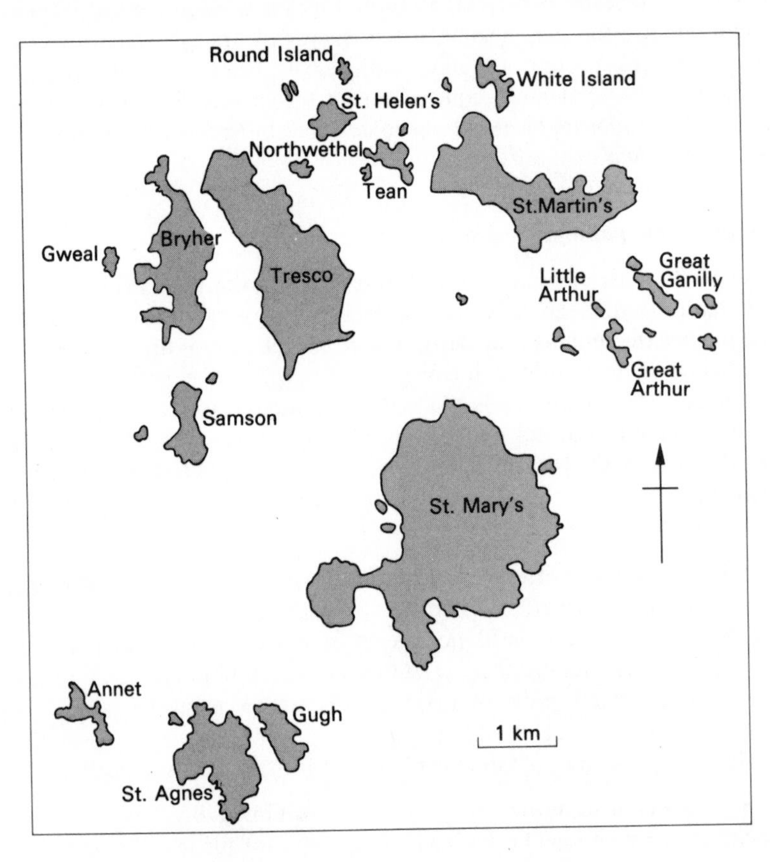

Figure 40. Isles of Scilly (smallest islands omitted)

prevailing. Small colonies, on the other hand, must frequently be faced with uncompromising and extreme forms of habitat and the necessity for rapid adjustment in order to avoid extinction.

The principle is beautifully illustrated by the spot-distribution in females of the meadow brown butterfly, *Maniola jurtina,* in the Isles of Scilly (Figure 40). Samples collected on the three largest islands, each with an area of 680 acres (270 ha) or more, exhibit a uniform kind of spotting with almost equal values at 0, 1, and 2 spots (Figure 41). The small amount of

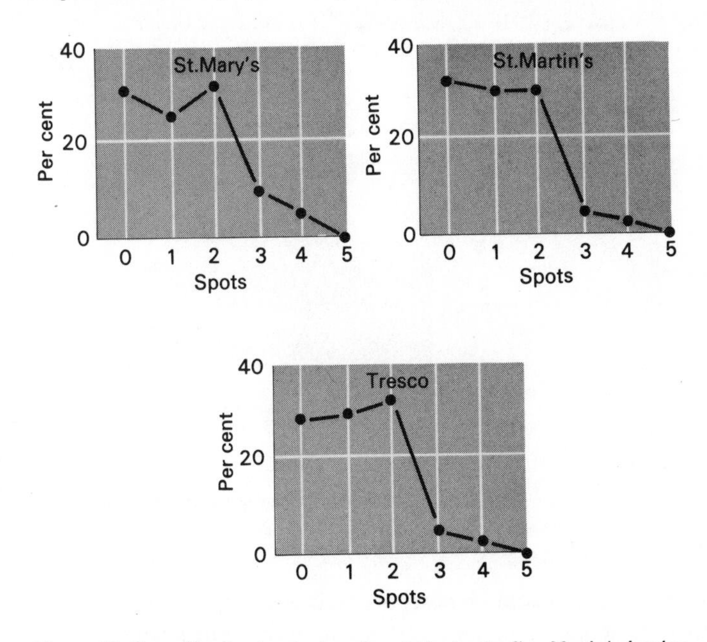

Figure 41. Spot-distribution in females of the butterfly, *Maniola jurtina,* on three of the largest islands in the Isles of Scilly. (From *Heredity.*)

spot-variation apparent from the diagrams is not statistically significant and the populations can therefore be regarded as identical in this respect. By contrast, the *jurtina* populations on the small islands, with areas of 40 acres (16 ha) or less, exhibit a great diversity of spot-values. Three typical examples are shown in Figure 42, and they obviously bear no resemblance to one another or to those on the large islands. Spot-values of the meadow brown in Scilly are apparently stabilized, and there is every reason to believe that this stability is due to the action of natural

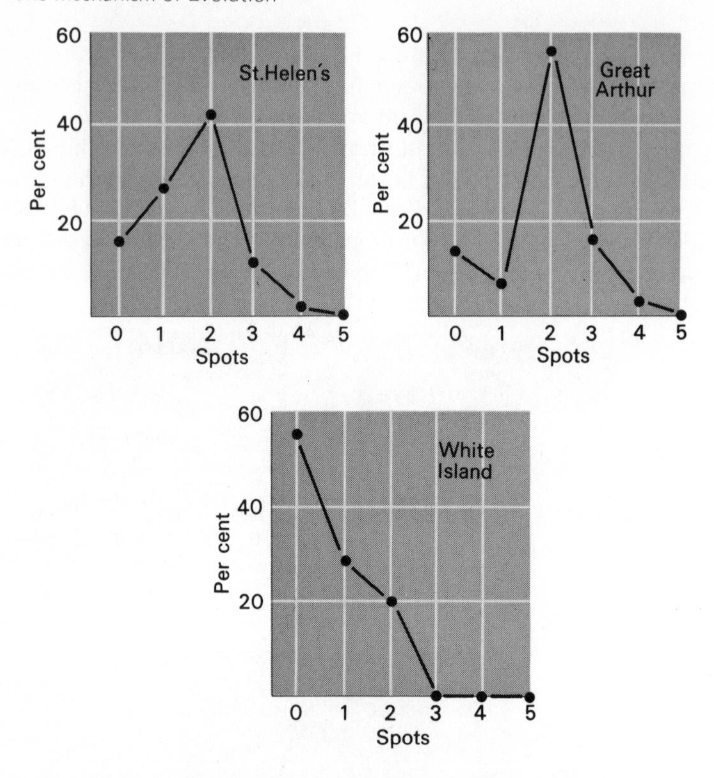

Figure 42. Spot-distribution in females of the butterfly, *Maniola jurtina*, on three of the small islands in the Isles of Scilly. (From *Heredity*.)

selection. The observed facts accord well with expectation, for it is likely that small, isolated populations will tend to become adapted to the special conditions prevailing in their restricted habitats. One of the most striking features of the small islands is the distinct ecology peculiar to each of them. The large islands, on the other hand, are all able to provide a similarly wide range of habitats and their populations of butterflies have evidently become adjusted to the average of the conditions that prevail there.

It could be argued that the flat-topped spot-distributions on the large islands are the result of combining samples containing a variety of spot patterns obtained from a number of closely associated yet distinct populations. This hypothesis was tested on one of the large islands (St. Martin's) by selecting three different collecting zones each with its own characteristic ecology. Comparison of the resulting samples gave $\chi^2_{(4)} = 3.68$ $(0.5 > P > 0.3)$, indicating homogeneity of spotting in the three localities and so discounting any suggestion that samples from distinct populations had been combined.

Random genetic drift

An alternative explanation of situations such as that just described has been put forward by Sewall Wright, who has pointed out the probability that mere chance may play a part in the survival and spread of particular genes, neutral in their effect under existing conditions. When considering the Hardy-Weinberg equilibrium (Chapter 2, page 29) we saw that if the frequencies of two alleles A and a are p and q, then in a stabilized situation the proportions of the three genotypes AA, Aa, and aa in the next and subsequent generations will be $p^2 : 2pq : q^2$. The equation only holds good, however, for large populations; it is untrue for small ones owing to the influence of chance, that is to say, of *random genetic drift*. That drift occurs is a mathematical certainty. The questions at issue are whether, by comparison with selection, its effects in promoting evolutionary change are significant, and if so, in what circumstances?

To take a hypothetical situation quoted by Berry, suppose two alleles, A and a, occur with equal frequency (i.e. $p = q = 0.5$) in two populations, one containing 500 000 individuals and the other 50. These will have been derived from the union 1 000 000 and 100 gametes respectively. In order to determine the number of A and a genes which will be withdrawn at random from a gene pool containing equal numbers, we must first calculate the *standard error* of each sample. For the large population this is,

$$\sqrt{\frac{500\,000 \times 500\,000}{1\,000\,000}} = \pm\ 500$$

That is to say, the true numbers of A and a genes lie within the range 500 000 ± 500, i.e. between 500 500 and 499 500.

A similar calculation for the small population shows the expected numbers of A and a alleles to be 50 ± 5, i.e. between 55 and 45.

Clearly, the proportions of A- and a-bearing gametes will vary much more from generation to generation in the small population than in the large one. Hence, for genetic drift to have an appreciable evolutionary effect by changing gene-frequencies, a population must be small. The larger a population becomes, the less will be the influence of drift because of,

(i) its decreasing contribution to overall variation,
(ii) the relative potency of natural selection.

Now the majority of the *jurtina* colonies we have studied have been 'large', containing thousands of individuals, but a few have been of the right order of magnitude for drift to be detectable. If it had been exerting an appreciable effect in these small communities we might have expected fluctuation in their spot-distribution from year to year. But this has not occurred. As we saw earlier, a feature of the populations in Scilly, both small and large, has been their great stability of spotting over a period of many years.

Attempts elsewhere to demonstrate the operation of random gene survival have also been rather unconvincing. On theoretical grounds there are various reasons why this should be so. For instance, it is well known that large populations as well as small ones are subject to changes in the proportions of different forms controlled by specific genes or gene-groups. These fluctuations are often of such size and occur so fast that they could never possibly be ascribed to random survival. Further, it now seems probable that most genes have multiple effects. A typical example has been analysed by Sheppard in the scarlet tiger moth, *Panaxia dominula* (page 92), where the *medionigra* gene has been shown to affect such diverse characteristics as mating behaviour, fertility, colour pattern, liability to attack by birds, and survival during the early larval stages. While it is possible that the adaptive value of a gene may be neutral within a limited set of circumstances, in a changing environment its effects are unlikely to remain neutral long enough for the effective operation of 'drift'.

The founder principle

Chance may also play a part in influencing the nature of populations through the erratic immigration of stragglers from other communities into localities such as islands, which are subject to a considerable degree of isolation. If the home population is large, the influence of fortuitous arrivals will be negligible, but if it is small or non-existent, the introduction of new alleles could have a profound effect on the gene-pool and hence on the range of variation available for selection. This so-called *founder principle* could thus operate significantly in circumstances where a number of localities existed, each differing in its ecology yet bearing a generalized resemblance to the others. Assuming that the colonizing immigrants were able to reach all the localities, this could lead to the formation of distinct races. However, from what has been said already, it will be realized that such situations are not necessarily explicable only through the reduction of existing populations to a few individuals or the arrival of founders. They could also be accounted for in other ways.

Polymorphism and race formation

The study of isolation in animals and plants frequently reveals instances of discontinuous distribution in which the same form of a species appears to occur in two or more distinct and isolated localities. Yet there is seldom any evidence by which we can judge whether discrete communities are simply offshoots of a single pre-existing one or have attained their similarity by independent means. Ford has shown that the existence of polymorphism can provide just the information necessary for the solution of this problem. The lesser yellow-underwing moth, *Triphaena comes,* is monomorphic in England, but in central and north Scotland, and also in the Scottish islands, a dark variety, *curtisii,* is found which is quite distinct in appearance from normal *comes* (Figure 43).

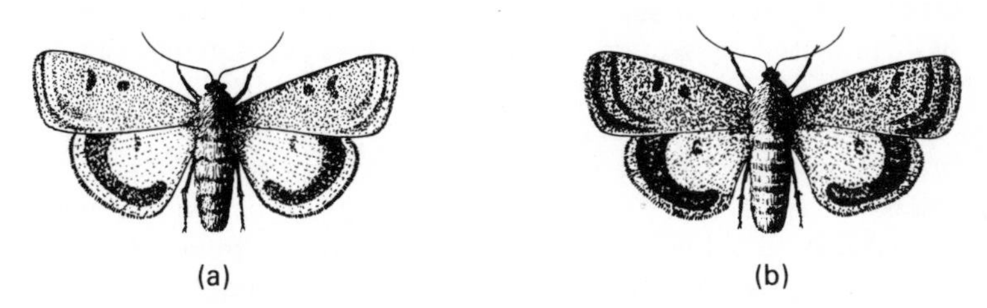

Figure 43. (a) Normal and (b) *curtisii* forms of the lesser yellow-underwing moth, *Triphaena comes*

The exact frequency with which these two occur in the wild state is not known, but the dark form is sufficiently common to suggest a balanced dimorphism. The colour of *T. c. comes* is remarkably constant and varies only slightly; that of *T. c. curtisii,* however, is very variable, the shade of the fore-wings ranging from mahogany to intense black, while the yellow hind-wings also tend to be tinged by dark scales or almost obscured by them. Breeding experiments with mainland insects have established that the difference between the two forms is due to a single pair of alleles, *curtisii* being nearly but not quite dominant to *comes.* The heterozygotes are thus all *curtisii* and are readily distinguishable from the lighter homozygous recessive *comes.* Between heterozygotes and homozygous dominants there is a good deal of overlap in colour; the darkest insects are nearly always homozygous and the lightest heterozygous, but intermediates may belong to either class. Stocks of *curtisii* were obtained from Barra (Outer Hebrides) and from Orkney, and crosses were made between them. The progeny of dark forms from the two localities were also dark, as was to be expected. But crossing heterozygous *curtisii* from Barra and Orkney resulted in a complete breakdown of dominance, and instead of a reasonably clear-cut segregation a continuously graded series was obtained, ranging from the lightest *comes* to darkest *curtisii.* This showed that although the same gene, or alleles of it, control *curtisii* on Barra and Orkney, the degree of dominance it achieves must be adjusted by other genes (modifiers), which are different on the two islands. Hence, although the dark forms are identical in appearance and genetic behaviour, their resemblance has been achieved by quite different means.

Viewed in wider aspect, one of the most striking contributions of polymorphism towards race formation is to be found among the human ABO blood groups, which may be controlled by a multiple allele series or, more probably, by groups of closely linked genes maintained by duplication (page 32). In Table 15 is summarized the distribution of this series among a number of different nations, showing the striking differences that exist between them.

Among the Hindus (at the foot of the table), the four groups evidently exhibit a balanced polymorphism, while at the other extreme, the frequency of the B group among the Australian aborigines (3.8 per cent) is almost low enough to be maintained by recurrent mutation; in effect, we therefore have a dimorphism of groups O and A. Not only do the ABO frequencies change from one country to another but they also tend to vary according to distinct patterns. Thus the value for group B increases as we go eastwards, ranging from 6 per cent in France, through 14.1 per cent in Germany, 22.2 per cent in Persia, to 37.2 per cent in India, where it reaches a maximum. Blood group distribution also throws an interesting light on the origin and affinities of isolated races such as the Basques in south-west France and the Hungarian gypsies (Table 15). The latter differ appreciably from the Hungarians with whom they are

Table 15. Percentage frequencies of the blood groups of the O, A, B series in different races (arranged in increasing frequency of group B). (After Ford.)

Race	Total examined	Frequency (per cent) of blood groups O	A	B	AB
Australian (aborig.)	603	54.3	40.9	3.8	1.0
Dutch	14 483	46.3	42.1	8.5	3.1
English (southern)	3449	43.5	44.7	8.6	3.2
Dutch Jews	705	42.[illegible]	39.4	13.4	4.5
Russian Jews	1475	36.6	41.7	15.5	6.1
Bushmen	336	83.0	–	17.0	–
Hungarians	1041	29.9	45.2	17.0	7.9
Arabs	2917	44.0	33.0	17.7	4.1
Japanese	24 672	31.1	36.7	22.7	9.5
Russians	57 122	32.9	35.6	23.2	8.1
Negroes (Congo)	500	45.6	22.2	24.2	8.0
Chinese (Canton)	500	45.5	22.6	25.0	6.1
Hungarian gypsies	925	28.5	26.6	35.3	9.6
Hindus	2357	30.2	24.5	37.2	8.1

closely associated, their nearest resemblance being to the Hindus from whom they may well have originated. Intermarriage and greatly increased mobility have inevitably tended to obscure human polymorphisms such as the blood groups. However, evidence that gene-flow can still be restricted in spite of the reduction of physical barriers is provided by the distribution of ABO in Britain, where the frequency of the O gene rises from about 0.67 in the south-east to 0.72 in Scotland and 0.74 in areas of Ireland.

From the examples of polymorphism discussed earlier, we might expect that the distribution of the human blood groups would be controlled by powerful selective forces. But it is only comparatively recently that this view has gained acceptance, partly because of a lack of understanding of the significance of polymorphism, and also due to the difficulty of pinpointing selective agents. However, evidence is now beginning to accumulate, still largely statistical, of relationships between particular blood groups and certain human conditions. Thus Aird, Bentall, and Roberts have demonstrated an association between cancer of the stomach and group A, where the incidence of the condition is up to 20 per cent higher than in other groups. Similarly, duodenal ulceration has been found to be between 17 per cent and 54 per cent commoner in group O than elsewhere.

For a selective agent to influence the composition of the gene-pool it must exert its effect during the reproductive period. One of the limitations in interpreting findings such as those outlined above is that many of the conditions, such as cancer and ulceration, are features of middle age, i.e. of post reproduction. Blood incompatibility between mother and foetus, on the other hand, could exert a strong selective influence and this can occur in ABO just as it does in the Rhesus group. The resulting condition, known as haemolytic disease of the new-born, shows an excess in O group mothers, suggesting that the disease is more likely to occur in a group O woman carrying a foetus whose ABO group is incompatible with her own, than in one who is either group A or group B. One of the most striking discoveries in this field has been made by Vogel and Chakravartti who have investigated the relationship in Indians between blood groups and smallpox. Not only are those in groups A and AB more liable to catch the disease than individuals of B or O, but they are also likely to be affected more severely. Mortality from smallpox was also found to be significantly higher in group A than in the rest.

Balanced polymorphism is as widespread in plants as it is in animals and leads equally readily to the formation of local races. But, here again, we still have comparatively little idea of the phenotypic effects of the genes concerned and hence of the precise nature of adaptation. Among plants, some of the best known polymorphic forms are concerned in controlling the method of reproduction, as in the primrose, *Primula vulgaris* (page 47). Others involve a wide range of structural and physiological characteristics many of which are closely interrelated. The small herbaceous plant *Hepatica triloba,* has white- and blue-flowered forms, both of which exist together in the same community. A small, well-colonized area near Michigan was divided into three zones, two shaded by a thick wood and the third a clearing exposed to the afternoon sun. The proportions of white-flowered plants present were found to be 5.67 and 6.54 per cent respectively in the shade, and 36.99 per cent in the lighter area. Similar experiments conducted under a variety of conditions confirmed the view that a significant relationship exists between the light and the frequency of the white form. In other species, such as *Ceanothus,* a comparable association appears to exist between climatic conditions, such as temperature and humidity, and the occurrence of flowers of a particular colour.

Occasionally it has been possible to relate the phenotypic effects of a polymorphism to the specific ecological requirements of a species, and hence to account for local race formation in more precise terms than is usually possible. An instance of this is provided by the clouded yellow butterfly, *Colias croceus* (Figure 44), an annual migrant to Britain from the

Continent, which is, however, unable to breed here. Numerous closely related species occur in North America, typical among them being the alfalfa butterfly, *C. eurytheme,* which has been extensively studied on account of its economic importance as a pest in the larval stage. In many of these species the males are monomorphic and yellow-coloured while the females are dimorphic, the commoner being yellow like the males while the rarer form is whitish in appearance. Light-coloured individuals usually amount to between 5 and 15 per cent of the females, although proportions as high as 50 per cent have been recorded by Hovanitz in North America. As far as is known, the female dimorphism is invariably controlled by a single pair of autosomal alleles *H* and *h,* that producing the white form being dominant (*H*) and expressing its effect only in the female gene-complex. A gene which exerts a phenotypic effect in one sex only is said to be *sex-controlled* (or sex-limited), not to be confused with a sex-linked gene (see page 35). The genetic situation is summarized in Table 16.

Table 16. Sex-controlled inheritance in the European clouded yellow butterfly, *Colias croceus*

Possible genetic constitution		
Male (yellow)	Yellow female (typical)	White female (*helice*)
hh or *Hh* or *HH*	*hh*	*Hh* or *HH*

Remington has shown experimentally that all three female genotypes can exist but their relative viability depends upon environmental conditions, particularly temperature. There can be little doubt that as far as the white forms are concerned, heterozygous advantage must play an important part in maintaining the dimorphism in the wild state (see also page 72). Under natural conditions, Hovanitz found that the difference in colour of the female is associated with a corresponding change in behaviour. Recording times of flight of the two forms, he found that the proportion of white was highest in the early morning and less towards noon; sometimes, but not always, it rose again in the late afternoon. The principal physical factors influencing female activity as a whole are radiation, temperature, and humidity, and it therefore seems likely that a combination of these is involved in controlling differentially the flight pattern of the two females. That of the white form appears to prolong the period of the insect's activity but to reduce its intensity, while in the yellow activity is concentrated towards the middle of the day. The balance between the two thus depends upon ecological considerations. As might be expected, the proportion of white females is found to increase with altitude and lower average temperature. Experimental studies have shown that the larvae of white females also exhibit a characteristic variation in that they develop more rapidly than those of yellow females. Again, this difference could be advantageous under cool conditions and lead to a restriction in gene-flow and the formation of local races.

Judging by a single set of results obtained in 1947 (Dowdeswell and Ford) there is a strong suggestion that the European clouded yellow, *Colias croceus,* and its white form, *helice,* behave in a manner similar to the North American species (Figure 44 and cover picture). Unfortunately, the insects were not sufficiently abundant that year for strict

Figure 44. Sex-controlled inheritance in the European clouded yellow butterfly, *Colias croceus.* White, *helice* female (*above*) and typical yellow female (*below*) (x 1)

Figure 45. Forms of the white-lipped snail, *Cepaea hortensis,* in their natural environment. To the human eye, banded and unbanded shells are equally visible against a grassy background (x ¾)

statistical significance to be achieved in the samples; nonetheless, the pale phase clearly showed a predominance in numbers at the beginning and end of the day.

Clines and gene-flow

We saw in Chapter 4 how the study of gradients of variation in a species (clines) can sometimes throw valuable light on the mode of formation of races and even of distinct species. One of the limitations of this approach is that investigations frequently need to extend over a distance of many miles, so that it is usually possible to gain no more than an approximate idea of the changing ecological conditions to which the organisms are responding. However, in circumstances where high selection pressures are operating in a diversified environment, we might expect to find relatively steep clines occurring over short distances of a few hundred yards. Such situations are now known to be quite common and, if studied intensively, can provide useful information on gene-flow and sympatric evolution. Day and I have investigated a population of the white-lipped snail, *Cepaea hortensis* (Figure 45), inhabiting grassy banks (the strips)

Figure 46. Grassy banks alternating with ploughed land on Portland Bill. The banks support large colonies of the white-lipped snail, *Cepaea hortensis*

on Portland Bill (Figure 46), a locality known to be isolated from neighbouring snail populations. Two adjacent banks about 360 m long were selected for study, and samples were collected from three 45 m stretches on each over a period of five years. The shells (all with yellow ground colour) were scored as banded or unbanded; these characters are known to be controlled by a pair of alleles, unbanded being dominant. The results are shown in diagrammatic form in Figure 47 from which it will be seen that a cline existed, with an average value at the west end of roughly 55 per cent banded and at the east end of about 20 per cent, with intermediate values between the ends. All samples collected were marked with dots of cellulose paint, thus enabling estimates of population numbers and movement to be made. The average population per 45 m stretch was about 800 (range 1800-400) while the average distance moved by a snail along a bank in a year was of the order of 6.4 m (maximum 85 m). Evidently, gene-flow within the area must have been considerable. In an attempt to isolate the selective agents involved in maintaining the cline, a detailed study was made of the two predators on adult snails–birds and

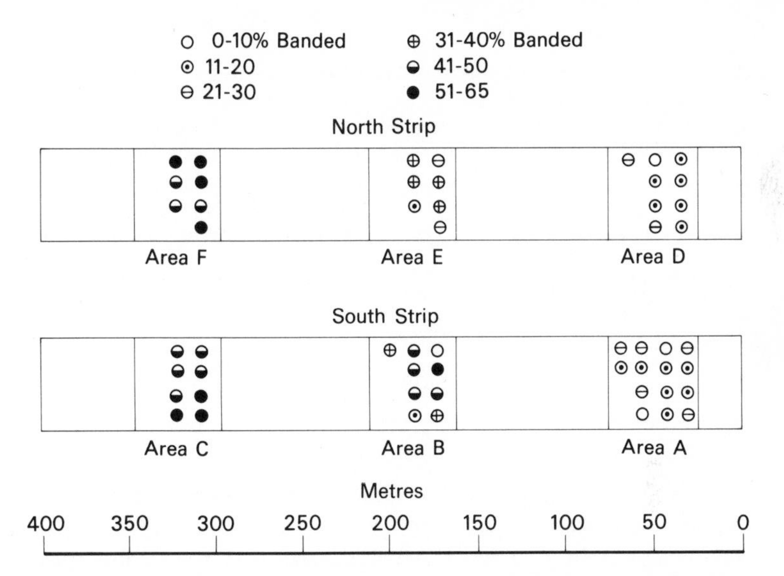

Figure 47. Variation in shell banding in two populations of the white-lipped snail, *Cepaea hortensis,* on Portland Bill, 1961-5 (From *Heredity*)

rodents. Cracked shells obtained from anvil stones (Figure 26, page 68) at the foot of the banks showed a consistent selection pressure by thrushes of about 8 per cent against banded shells, irrespective of whether they had been withdrawn from high or low banded populations. However, the level of predation in relation to the total population was relatively low throughout four years, averaging about 4.2 per cent overall (maximum 12.8 per cent for areas A, B, and C in 1962). As would be expected, rodent predation (which takes place at night) showed no selection, the proportion of banded individuals among cracked shells being that of the living population. The situation in *C. hortensis* inhabiting the strips has remained constant over a period of ten years and it is clear that powerful forces must be acting to maintain the cline in the face of considerable gene-flow. Equally, it seems unlikely that the low level of selective predation by birds could account for more than a small proportion of the stabilization. To account for it completely would have necessitated differential selection along the length of each bank, for which there is no evidence. Other possible selective agents are not easy to suggest as the ecology of the area consists of remarkably uniform grassland, the only obvious variation being a drop of about 1.2 m between the west and east ends of each bank.

Evidence that more subtle influences are at work is provided by samples of young snails collected from the same area (Table 17).

Table 17. Comparison of different ages in *Cepaea hortensis* from Portland Bill with respect to banding. (From Heredity.)

Date	Shells over 10 mm diam.			Shells 10 mm diam. and less		
	Unbanded	Banded	Total	Unbanded	Banded	Total
12-13.8.67	57	12	69	19	4	23
7-8.10.67	124	27	151	18	—	18
16-17.8.68	100	31	131	35	4	39
Total	281	70	351 (19.9% banded)	72	8	80 (10.0% banded)

Comparison of the two sets of data gives $\chi^2_{(1)} = 4.35$ $(0.05 > P > 0.02)$ indicating that banding was at an advantage in the older snails compared with the young ones (the reverse of the situation resulting from bird predation). Again, the nature of the selective agents involved is at present unknown, but they do not include birds. The cline of banding existing in the adult snail population could well be maintained by a balance of forces acting at different stages in the life cycle, the point of equilibrium between them varying according to a gradient of ecological requirements.

Area effects

Situations have been described by Cain and Currey in *Cepaea*, where particular phenotypes may predominate in certain diversified localities irrespective of visual selection by predators. Such a constancy of morph frequency is known as an *area effect*, and may cover a zone ranging from several square miles to a few hundred square yards. Comparable situations are also known in plants, for instance in the cyanogenic forms of *Trifolium* and *Lotus* (see page 125). It seems unlikely that over a wide and continuous area such as chalk downland, the origin of each of these distinct populations could be ascribed to founder individuals (page 100). Moreover, since the numbers involved are usually large, genetic drift will have exerted only a minimal effect in establishing particular gene-frequencies. The most likely explanation of area effects would seem to be that of the various polymorphic systems available, each adjusts physiologically to those ecological conditions in which it is at a maximum selective advantage. Where two different stabilizations impinge on one another, a steep cline for a particular variant may develop between them. Such a happening in the past could account for the cline of banding in the strips population of *Cepaea hortensis* on Portland, described in the previous section. In this instance, the two original populations could have been subsequently eliminated by ploughing, leaving behind only the bridges between them with the high differential selection pressures still in force at each end.

Selective advantages and the rate of evolution

Numerous attempts were made during the early years of this century to demonstrate under experimental conditions the survival-value of particular characteristics in wild organisms, such as the advantage of different degrees of protective coloration in insects attacked by the same predator. The results were generally disappointing and seldom justified the conclusions reached from mere observation. We now know that this was only to be expected, for it is relatively seldom that a particular combination of genes will confer such benefits on an organism that it can sweep through a population in the course of a few generations. However, under the influence of man the ecological environment can change very rapidly, and we have only to look at the Midlands to see the effects of pollution that have resulted from the Industrial Revolution. As we saw in Chapter 3 (page 43), the blackening of trees with soot and the disappearance of epiphytes such as lichens which normally grow on their bark, has favoured the spread of melanic forms of moths such as the scalloped hazel, *Gonodontis bidentata* (Figure 16), which rely on concealment as a means of avoiding predation by birds.

One such species, the peppered moth, *Biston betularia,* has been studied extensively; it possesses a black form *carbonaria* (Figure 48(a) and (b)) which, like that of *G. bidentata,* is under the control of a single pair of alleles, that causing melanism being dominant. The first specimen was recorded from Manchester in 1850; now, more than 95 per cent of the population is of the black form. The remarkable spread of *carbonaria* in the course of a century has been studied by Kettlewell, who has shown that it has invaded rural districts as the level of pollution rose. Thus, the proportion of melanics in Winchester in 1962 was about 14 per cent. Kettlewell's beautifully designed experiments have shown conclusively that the selective agents involved are birds such as the hedge sparrow, spotted flycatcher and nuthatch. The reason why predation had never previously been observed was that a bird seizes the resting moth too rapidly for the action to be detected except by a carefully planned

Figure 48(a). Typical and melanic (*carbonaria*) forms of the peppered moth, *Biston betularia,* at rest on a soot blackened tree trunk (x 1)

experiment involving many moths and numerous observers. The typical insect is almost invisible to the human eye as it rests on lichen-covered trees in rural districts. Similarly, on the soot-blackened vegetation of industrial areas the black form is equally well concealed (Figure 48(a) and (b)). By releasing equal numbers of the typical and melanic forms of *B. betularia* on to tree trunks in a rural area and observing the action of birds from hides, Kettlewell was able to demonstrate differential selection by predation. In all some 190 moths were eaten by insectivorous birds, 164 of them being *carbonaria* and only 26 typical. Kettlewell also released large numbers of moths, both melanic and typical, in two localities, one unpolluted (Dorset) and the other polluted

Figure 48(b). Typical and melanic (*carbonaria*) forms of the peppered moth, *Biston betularia,* on an unpolluted tree with a growth of lichen (x 1)

(Birmingham). Each specimen was marked with a dot of cellulose paint for identification, and samples from the experimental populations were collected periodically at night by attraction to assembly bags containing freshly emerged females and to mercury vapour lamps. For this purpose it was necessary to use males only as females are not attracted to males by scent nor do they tend to fly towards light. The numbers released and recaptured are summarized in Table 18.

As will be seen from the table, in the unpolluted area of Dorset recaptures of typicals were about double those of *carbonaria,* while in the polluted Birmingham locality the situation was reversed, the black form being at an advantage.

Table 18. Numbers of peppered moths, *Biston betularia,* liberated and recaptured in two different localities. (After Kettlewell.)

Locality		Typical	*carbonaria*	Total
Dorset 1955 (unpolluted)	Liberated	496	473	969
	Recaptured	62	30	90
	% recaptured	12.5	6.3	
Birmingham 1953 (polluted)	Liberated	137	447	584
	Recaptured	18	123	141
	% recaptured	13.1	27.5	
Birmingham 1955 (polluted)	Liberated	64	154	218
	Recaptured	16	82	98
	% recaptured	25.0	52.3	

Important confirmation of these findings has been provided by the work of Clarke and Sheppard, who have conducted a limited survey of melanic *B. betularia* with particular reference to its distribution in the Liverpool area, and the effect of smokeless zones with the reduced pollution resulting from them. Moving westwards in the direction of the prevailing wind, a rapid decline in the incidence of *carbonaria* was found to occur from a value of about 97 per cent in industrial Liverpool to less than 10 per cent in the country fifty miles further west (North Wales). In one locality (Caldy) the frequency of *carbonaria* had dropped from a maximum of 94.2 per cent in 1960 to 90.2 per cent in 1965, perhaps as a result of the introduction of smokeless zones in the vicinity. A comparable situation has also been reported in 1970 from Manchester. Experiments involving the fixing of dead moths to tree trunks in life-like positions both at Liverpool and Caldy, have confirmed the view that the melanic form is at an advantage relative to bird predation when in sooty surroundings. In such circumstances (Liverpool) the selective disadvantage of the typical form appeared to be of the order of 60 per cent. At Caldy the disadvantage had previously been about 50 per cent, but with the introduction of the smokeless zone this had fallen to around 20 per cent in the course of five years.

Haldane suggested that the gene-frequency of the black moths in the Manchester area (that is to say the number of *carbonaria* genes *present* in the population expressed as a percentage of the total number of loci *available* for them to occupy), may have been about 1 per cent in 1850. Latterly it must have reached nearly 99 per cent, an advance which could have necessitated a 30 per cent selective advantage of melanics over typical individuals.

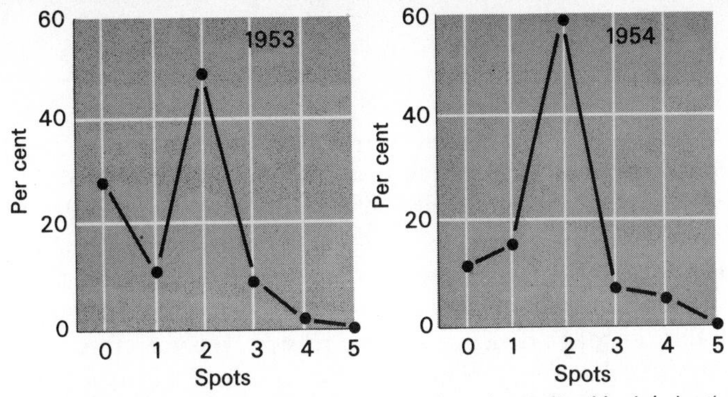

Figure 49. Spot-distribution in females of the butterfly, *Maniola jurtina,* on the western area of Tean, Isles of Scilly in 1953 and 1954, showing the change associated with altered ecology. (From *Heredity.*)

The realization that selective advantages of 30 per cent or more are quite common among polymorphic species, instead of rarely exceeding 1 per cent as had previously been supposed, represents a major change in evolutionary thought during recent years, and we have already seen how industrialization promoted the rapid spread of melanic peppered moths in the course of a century.

Under the most powerful conditions of selection we might expect that the process of evolution would be detectable over a period of a few years. Ford and I have shown that this is indeed so and that appreciable changes can take place within the gene-pool of a population in a single generation. On the small island of Tean (Isles of Scilly) grazing by a herd of cattle had kept the vegetation consistently cropped and the grassy areas almost lawn-like. The island supported a large population of the meadow brown butterfly, *Maniola jurtina,* to which the 'lawns' proved to be a physical barrier on account of their wind-swept nature, restricting gene-flow across them almost completely. The removal of the cattle resulted three years later in a rank growth of vegetation, areas which had once been an expanse of short grass assuming the appearance of a hay field with a considerable subsidiary vegetation of bramble and bracken. These changes enabled a colony of the butterfly to spread from a small, isolated locality into an adjoining one which now afforded the necessary protection from the wind, whereas it had not previously done so. The resulting adjustment in spotting of the females over the period of a year between 1953 and 1954 is shown in Figure 49. Evidently, a significant evolutionary change had occurred in a single generation, from a bimodal spot-distribution (modes at 0 and 2 spots) to a unimodal one (at 2 spots). The difference is given by $\chi^2_{(3)} = 8.8$ ($0.05 > P > 0.02$). These findings incidentally serve to emphasize the point discussed earlier (Chapter 3), namely that the nearer a species is to the periphery of its range (*M. jurtina* does not extend further west than the Scillies) the more sensitive does it tend to become to small environmental changes.

In order to calculate the selective advantage or disadvantage enjoyed by a particular gene we must first estimate the gene-frequency (see page 114). Ideally, this would involve counting the whole population and determining the proportion of the appropriate phenotype present, but in practice this cannot be done for obvious reasons. It is therefore necessary to employ some sampling technique such as is used, for instance, in the capture-recapture method (see Chapter 6). The work of Fisher, Ford, and Sheppard on the colony of the scarlet tiger moth, *Panaxia dominula,* at Cothill, Berkshire, has demonstrated admirably the great possibilities of this procedure. The insect is day-flying, the fore-wings being a metallic-green colour with white and yellow markings, while the hind-wings are scarlet and black. When at rest the fore-wings only are visible, and the moth is surprisingly well concealed in its natural surroundings. In flight the movement of the hind-wings gives a flashing effect, making the course of the insect difficult to follow. The bright colours are probably aposematic (warning), being exposed when the insect is alarmed at rest. At the same time, two drops of amber-coloured fluid are exuded, one from each side of the prothorax. The liquid is not unpleasant to human beings, but available evidence suggests that the moth is seldom attacked by such potential predators as birds and dragonflies. *P. dominula* is polymorphic (Figure 50) and therefore particularly suitable for study. The expression of a single gene (*M*) showing no dominance (the normal allele being *D*) can be detected in the heterozygote (*DM*) by the reduction or absence of some of the markings on the fore-wings and a characteristic change in the black pattern on the hind-wings; this form is known as variety *medionigra.* In the rare homozygote (*MM*), called *bimacula,* a great extension of the black pigment obscures most of the red on the hind-wings and the majority of

Figure 50. Polymorphism in the scarlet tiger moth, *Panaxia dominula.* The letters in brackets refer to the genetic constitution of each form

Table 19. Frequency of *medionigra* in a population of the scarlet tiger moth, *Panaxia dominula,* at Cothill, Berkshire. (After Ford.)

Year	*dominula*	*medionigra*	*bimacula*	Total	Gene frequency (%)
1939	184	37	2	223	9.2
1940	92	24	1	117	11.1
1941	400	59	2	461	6.8
1942	183	22	0	205	5.4
1943	239	30	0	269	5.6
1944	452	43	1	496	4.5
1945	326	44	2	372	6.5
1946	905	78	3	986	4.3
1947	1244	94	3	1341	3.7
1948	898	67	1	966	3.6
1949	479	29	0	508	2.9
1950	1106	88	0	1194	3.7
1951	552	29	0	581	2.5
1952	1414	106	1	1521	3.6

the spots on the fore-wings. It is a peculiar fact that these polymorphic forms are unknown in any other colony except Cothill (only one or two specimens of *medionigra* have ever been recorded elsewhere), which suggests either that the mutation rate of the *medionigra* gene is exceptionally low, or that selection is more adverse in other localities.

Since all three genotypes are easily distinguishable at sight, it has been possible to calculate the gene-frequency yearly since 1939. A portion of these data is summarized in Table 19, from which it will be seen that over a period of fourteen years the frequency of the *medionigra* gene fluctuated from a minimum of 2.5 per cent in 1951 to a maximum of 11.1 per cent in 1940. The general trend subsequently has been towards a gradual reduction in *medionigra,* and in 1968 its gene-frequency was estimated at only 1.1 per cent. Sheppard has shown that over the period 1939-1952 the heterozygotes were fairly consistently at an average selective disadvantage of about 8 per cent compared with normal *dominula.* Examination of specimens caught at Cothill prior to 1928 has suggested that the gene was then probably at a low frequency of about 1.2 per cent or less. Thereafter, until 1939, it must have spread at a great rate, corresponding to a selective advantage of some 20 per cent.

Field observations provide little evidence regarding possible reasons for such changes, and it is not clear which characteristics controlled by the *medionigra* gene proved so beneficial. However, as we saw earlier (page 92), preferential mating in which a *dominula* female tends to select a *medionigra* male rather than one of her own kind, would tend to increase the gene-frequency in the population. Conversely, laboratory breeding experiments have shown reduced fertility among *medionigra* males compared with *dominula*, and also that its expectation of survival from egg to imago is only 75 per cent of that of the typical form.

Another important outcome of the work on *dominula* has been the accumulation of the data necessary for a critical examination of the possibilities of genetic drift (page 99). Since 1941 the size of the population has been estimated annually by the capture-recapture method (page 122), and throughout most of that time it has fluctuated between 1000 and 1800. However, from 1962 onwards the population tended to be small, and the expected number of *medionigra* must have been of the order of seven or less on several occasions. In spite of this, the influence of random genetic drift did not bring about the disappearance of the gene, as might have been predicted. Since the gene-frequency of *medionigra* is known for the whole period of thirty years, it has been possible for the first time to make a direct comparison of the potential effects of drift and selection in a wild population, and to calculate whether or not the observed fluctuations were sufficiently small to be explained by Sewall Wright's theory. The results show clearly that random gene fluctuations, such as might occur in very small colonies, could not possibly account for changes of the magnitude observed at Cothill. The variations in gene-frequency were far too large even during periods when the population was at its lowest density. Moreover, if the influence of drift had been appreciable, we might have expected the selective values of the three forms to have fluctuated during the period shown in Table 19. In fact they remained remarkably stable.

P.dominula is a distinctly variable species, and in *medionigra* one such variant is an additional black spot on the red hindwings, which is not present in the typical form. Only three generations of selective breeding in the laboratory were found to be necessary on the one hand to intensify the expression of the gene causing enlargement of the black spot to a patch as in *dominula*, and on the other to reduce it almost completely. The laboratory findings have now been beautifully replicated in wild populations, for at Cothill the size of the *medionigra* spot has gradually increased until it is now, on average, about double its original size, while that at Sheepstead Hurst (a nearby population into which *medionigra* was introduced in 1954) has been reduced to almost total recessiveness. As Ford and Sheppard have pointed out, this is, perhaps, the first time that a laboratory experiment has forestalled evolutionary changes in nature.

6

Experimental Study of Evolution

The previous account has, I hope, shown that the complex process of evolution can be studied experimentally in a variety of ways and that the experimental approach is the one most likely to provide a realistic idea of the magnitude and nature of evolutionary changes taking place in wild populations. The essence of success in any experiment is to ask the right questions; that is to say, the design must be such that the data obtained are susceptible of analysis in a *pre-determined* way. So frequently one hears of instances where vast amounts of numerical information have been laboriously collected, sometimes for many years, which might have been invaluable in elucidating evolutionary trends in a particular locality or species, but for the neglect of one vital aspect (often easily studied), which would have made statistical analysis possible.

This warning applies particularly to Natural History Societies, which are often well placed for keeping long-term routine records of animal and plant communities. Statistical advice can be obtained from all university biology departments and from most government and industrial research organizations as well; there are also a number of good books on statistical methods.

The types of information required for the study of evolution vary greatly, but in general it is probably true to say that four kinds of data are needed more often than any others. These concern,

(i) gene-flow within populations and between them,
(ii) population numbers,
(iii) selection: its magnitude and the agents concerned,
(iv) adaptation of organisms to their environment.

One of the difficulties in studying animals arises from their ability to move about. But if we are to study their evolution it is essential that we should know the extent of this movement, and whether we are dealing with a single large population or a number of smaller ones. In recent years we have become more aware of the significance of minor ecological barriers in carving up a community into a number of small and biologically distinct parts. The influence of restricted movement leading to an interruption in gene-flow has now been demonstrated in numerous species, an example being the meadow brown butterfly, *Maniola jurtina,* in the south of England and the Isles of Scilly. Here it has been shown that the minimum barrier necessary for the formation of distinct races is 230 m of wind-swept ground devoid of vegetation more than about 7.5 cm high. Furthermore, detailed studies of plant and animal populations have established beyond doubt that where selection pressures are high, a particular variant can assume different values within the same population in the absence of any ecological barriers (sympatric evolution).

Marking methods

The chief requirement for the study of dispersal in plants and animals is a method of marking individuals at various stages of their life cycle. Adult plants (other than floating forms) are seldom dispersed as such, and their distribution on the ground can generally be mapped by eye. Animals, on the other hand, must be marked in order to ensure identification. This can be done in a variety of ways, such as the use of metal rings or clipping fur in mammals, ringing of birds, attaching metal tags to the fins of fishes, or the use of dots of quickly drying cellulose paint on the wings or bodies of arthropods and the shells of molluscs.

Marking young stages such as eggs, seeds or spores until they reach adult form presents greater problems, owing to the transformations which most animals and plants undergo during their early life. None of the methods enumerated above would be suitable for this purpose, but radioactive isotopes have great possibilities. A typical example of their successful use is provided by Kettlewell's quantitative study of moths by marking the larvae with radioactive sulphur. In order to be suitable for this sort of work, an isotope must have a half-life long enough to survive the pupa stage yet sufficiently short to avoid prolonged contamination of the countryside. Moreover, it must be easily ingested and incorporated both by the food plant (through which it is introduced) and by the insect itself. Sulphur-35 fulfils all these requirements, having a half-life of 87.1 days and being a constituent of many plant and animal proteins. The species used by Kettlewell were the scarlet and garden tiger moths, *Panaxia dominula* and *Arctia caja,* and their food plants, dead nettle (*Lamium* spp.) and dock (*Rumex* spp.). The larvae were fed for sixty-five hours on food which had been stood for two or three days in a solution containing 100 micro-Curie of sulphur-35 per litre. Larvae of *dominula* subsequently presented at the mica window of a Geiger-Müller counting tube were found to be radioactive, and their excreta even more so. When the adults emerged six to eight weeks later, not only were they, too, radioactive but their counts were far in excess of those obtained during the larval and pupal stages. This was no doubt due to the increased surface area of the imago and to the soft rays produced by the sulpur-35 being largely self-absorbed in the cylindrical larvae and pupae.

Work on *A. caja* confirmed these results and also showed that eggs may give higher counts than the radioactive females which laid them. The survival rate of treated specimens compared with controls gave no evidence that the presence of sulphur-35 is deleterious to the insect at any stage of its life history, so that here we have a method of marking with great possibilities for the study of evolution.

We have already considered one example (that of *Panaxia dominula*) in which it has been possible to detect differential survival in three polymorphic forms of a wild species and to treat these quantitatively (page 116). In this instance the phenotypes were easily recognizable so that counting presented no difficulty. The use of radioactive tracers such as that just described has opened up many new lines of approach which have previously been beyond us, for example the study under natural conditions of the action of genes whose only known effects are physiological. Moreover, it should now be possible to follow the success of particular genotypes in the wild state throughout the whole of an organism's life cycle.

Estimating populations

The estimation of numbers in plants presents relatively few difficulties from a practical standpoint. It will seldom be possible to count the individuals in a complete population, so that some method of sampling must be adopted. This generally involves the use of quadrats—unit areas of a convenient size (often 1 square metre) selected at random within the colony in question. Some of the less mobile animals, such as molluscs, can be sampled in a similar way. Small organisms inhabiting soil or water can be extracted by a variety of means from unit volumes of the medium in which they occur. Aquatic forms can often be filtered, while many small soil arthropods are negatively thermotactic and can therefore be removed by heating. Earthworm populations are often sampled by chemical methods, such as pouring a solution of formaldehyde on the soil, which causes the worms to come to the surface. A particular problem is presented by swiftly moving species such as many insects and most vertebrates. For these the capture-recapture method has proved extremely valuable as a means of estimating the numbers of a population.

Suppose we are dealing with a colony of day-flying moths such as *Panaxia dominula,* already described, and we catch 100 specimens and mark them so that they can be easily identified if recaught. They are then released and allowed to assort themselves once more within the population. In a further sample of 100 taken the next day, 10 are found to be marked. We can then calculate the total flying population on the first day as

$$\frac{100 \times 100}{10} = 1000$$

The principal limitations of this simple method will be obvious enough, namely, that it can only be employed with animals which randomize quickly once released, and in species where ability to learn plays no part in influencing the composition of the samples obtained. This latter reason renders the procedure rather unsuitable for birds and mammals, which can only be sampled by trapping. Some individuals quickly become trap-shy, while others associate capture with food and are caught with great regularity.

Used under suitable conditions, the capture-recapture method has proved extremely consistent and reliable. Results obtained over a period of sampling can be conveniently tabulated in the form of a triangular trellis (Figure 51) with the dates of samples running horizontally along the top. From each of these, lines run downwards south-east and south-west at 45 degrees so as to intersect. The total daily captures are entered at the end of the column running south-west from the date in question, while the total insects released are shown at the end of the corresponding line running south-east.

These two numbers will normally be the same and will differ only in the event of an insect dying or becoming damaged between the time of capture and subsequent release. From Figure 51 it will be seen that this evidently occurred on 22 August when, of twelve butterflies captured, only eleven were liberated. A dash (–) shows that no recaptures were possible on a particular day, either because no insects were caught or because none were released on the previous day.

Two examples will make this method of presentation clear. In

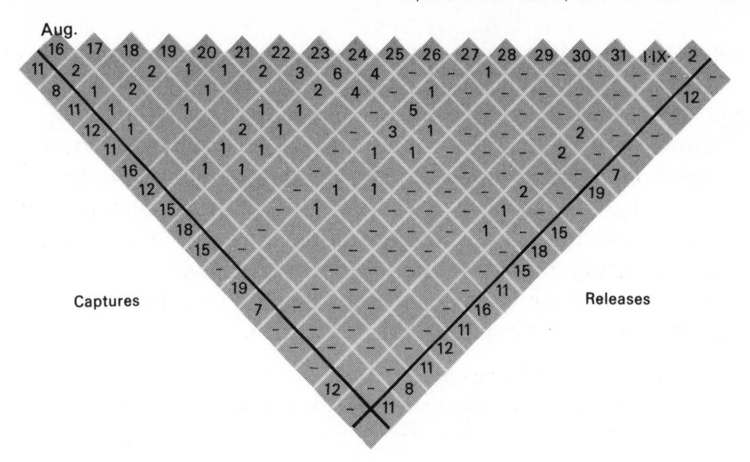

Figure 51. Results tabulated as a trellis in the capture-recapture method of estimating animal populations. Data obtained from a population of the meadow brown butterfly, *Maniola jurtina*

Figure 51 at the extreme left of the table and below the date 16 August is the entry 11 at the head of the 'Captures' column. A corresponding figure appears under 'Releases' on that date, showing that the whole sample was successfully marked and liberated. On the next day eight insects were caught, among them two captured the day before (these are entered in the body of the table in the square to the south-west of the appropriate date). Those not previously caught will each bear one mark only when released, while the recaptures will have two, one for each date. All eight captures were successfully released, and the entry appears in the space corresponding to 17 August. Again, the entries in the columns for 26 August are all dashes, indicating that no sample was taken that day and there were therefore no recaptures.

Direct estimates can be made of the daily flying population by the method already explained. For example, consider the total captures from 24 August onwards: out of 15, 19, 7, and 12 insects caught on subsequent dates, 4, 5, 1, and 1 respectively belong to the 18 marked on 24 August, i.e. 11 out of 53. Thus the total number flying on that day was about

$$\frac{53 \times 18}{11} = 87$$

The use of this simple procedure depends on the assumption that the population density remains constant from day to day. In fact, this will seldom be so, for immigration and emergence (in insects) will tend to increase it, while emigration and deaths will have the reverse effect. The extent of these fluctuations can be estimated mathematically by a comparison of the recaptures *expected* with those actually *obtained*, using as a basis the number of marks existing in the population from day to day. Space does not permit of further treatment here; for details reference should be made to the appropriate literature.

The capture-recapture method is also applicable to larval populations marked by radioactive sulphur (page 120). Thus,

$$\text{total larval population} = \frac{\text{radioactive larvae released} \times \text{total adults caught}}{\text{radioactive adults caught}}$$

Once we know the size of the larval population, and also that of the adults (obtained by the capture-recapture method just described), we can obtain an estimate of larval mortality by subtracting one from the other. Such techniques used for an appropriate period of time, enable us to compare the rates of survival of different forms, such as those of *Panaxia dominula* (page 116), and so to detect and measure the action of natural selection.

The efficiency of capture-recapture calculations has previously been limited to some extent by the laborious mathematical procedure involved. As might be expected, the advent of the digital computer has transformed the scene and made possible a number of refinements. For instance, it is now a simple matter to calculate the error involved in estimates of rates of survival and hence to gain a much more precise idea of the magnitude of selection pressures operating in a species under differing environmental conditions.

Adaptive significance of variations

The problems of studying natural selection in its varying aspects and the adaptations of organisms to their environment require little amplification, since they have provided many of the examples in the preceding chapters. Suffice it to add here that with the advent of more sophisticated biochemical techniques such as the electrophoretic separation of enzyme systems, a whole range of chemical polymorphisms has been revealed in a number of organisms, including man, of which we were previously unaware. Although we are still far from understanding their adaptive significance, the presence of these polymorphisms has served to emphasize the point that, being the products of genes, all adaptations are ultimately biochemical in origin. The principle is well illustrated by the studies of Daday and others of the white clover, *Trifolium repens*, and allied plants such as the birdsfoot trefoil, *Lotus corniculatus*, which exhibit a dimorphism for cyanogenesis. This involves the production of detectable quantities of hydrogen cyanide due to the interaction of a β-glucosidase (linamarase) with two cyanogenic glucosides, linamarin and lotaustralin. The basic hydrolysis reaction can be represented as follows:

$$\underset{\text{(glucoside-lotaustralin)}}{(C_2H_5)(CH_3)C(C_6H_{12}O_6)(CN)} \xrightarrow{\text{(enzyme-linamarase)}} \underset{\text{(ketone)}}{(C_2H_5)(CH_3)C{=}O} + HCN + C_6H_{12}O_6$$

In a normal leaf, glucoside and enzyme must somehow be kept apart, but it is not known how this is achieved. Should a leaf be damaged, the presence of cyanide can be detected in a few hours by sodium picrate paper which turns from yellow to reddish brown—a test which can easily be applied in the field.

Cyanogenesis in *Trifolium* and *Lotus* is controlled genetically by two pairs of alleles, the presence or absence of glucoside being determined by genes *Ac* (presence) and *ac.* Similarly the dominant *Li* determines the presence of enzyme and the recessive homozygote *li li* its deficiency. The reactions of the four genotypes in relation to cyanide production are summarized in Table 20.

Table 20. Reactions of the four genotypic combinations in white clover, *Trifolium repens,* with sodium picrate paper

Genotype	Reaction with sodium picrate paper
Ac Li	Rapid and strong
Ac li li	Slow and weak
ac ac Li	None, unless glucoside added
ac ac li li	None, even when mixed with either enzyme or glucoside

The adaptive significance of cyanogenesis is of considerable interest but is still open to a good deal of speculation. While there seems little doubt that in *Lotus* the capacity to produce cyanide confers on the plant a considerable degree of immunity from attack by herbivores such as molluscs (slugs and snails), in *Trifolium* this attribute is more doubtful and requires further investigation. Moreover, as Crawford-Sidebotham has shown, different species of Gastropods vary greatly in the degree to which they exhibit differential eating. Studies by Daday of the distribution of *Trifolium* in relation to temperature have demonstrated a

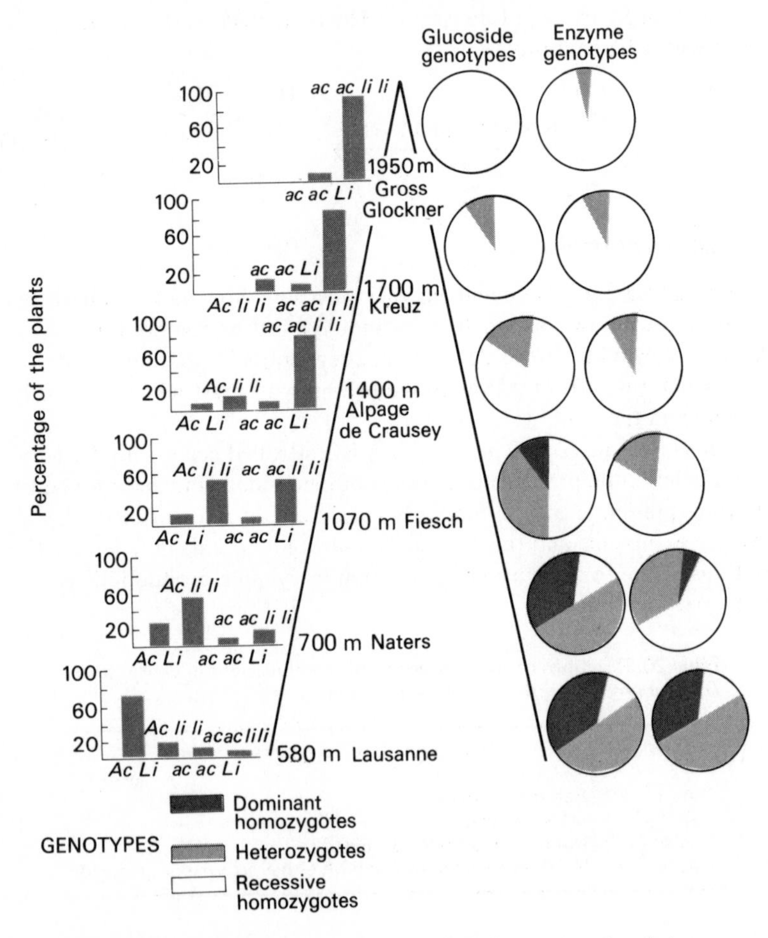

Figure 52. Cyanogenesis in white clover, *Trifolium repens,* in relation to altitude. (After Daday.)

close correlation between the incidence of cyanogenesis and both the January isotherm and altitude. Evidence for the latter in the Alps is provided in Figure 52, where an increase in altitude of little more than 1000 metres is associated with a total disappearance of cyanogenic plants. Whether the failure to adapt to cold conditions is a phenotypic effect of the *Ac Li* genotype or of closely linked alleles having other effects is not known. Suffice it to add that under laboratory conditions, refrigeration at 0 °C or below for about half an hour can cause the production of cyanide and the death of the leaves concerned. Arguing along similar lines, namely that edaphic factors may be exerting powerful selective effects, Foulds and Grime have shown, under experimental conditions of severe drought, that the rate of survival among *Ac* phenotypes of *Trifolium repens* is three times lower than that of *ac*. However, in moist conditions *ac* plants attained higher yields than *Ac*, suggesting a possible linkage of the *Ac* gene with others affecting vegetative vigour. A similar lack of vigour was shown by cyanogenic plants under normal conditions in relation to sexual reproduction, while a reduction in moisture inhibited their flowering altogether. In the field, Jones has found certain localities in the Netherlands where the frequency of cyanogenic plants in a population of *Lotus* changes from 20 to 80 per cent over a distance of less than 5 km–a cline which might well have arisen from edaphic influences.

The situation in *Trifolium* and *Lotus* thus provides an example of two balanced polymorphisms, one being related to physical factors while a second is concerned with predation (particularly in *Lotus*). However, the field is still wide open for the further study of cyanogenesis in all its aspects, not only in the two plant species mentioned here but over a wide range of other species as well.

7

Some Further Problems in Evolutionary Theory

In the previous chapters, a number of fundamental issues such as variation, adaptation, selection, and polymorphism have predominated, and these have been considered in some detail. Other important ideas such as the species concept and evolution above the species level have also recurred briefly at various points, and it may be profitable in conclusion to review these more broadly in the light of earlier findings. We may also legitimately ask whether the whole of organic evolution, as we know it, can be accounted for through a neo-Darwinian mechanism of the kind already described.

The species concept

In Chapter 1 we examined the views of the pre-Darwinian systematists on the nature of species and noted the fact that they made use mainly of structural resemblances in grouping organisms together. It is sometimes claimed as paradoxical that although Darwin called his famous book *The Origin of Species* (in its abbreviated title), nowhere in it does he seek to provide a satisfactory definition of a species. The argument in favour of such an omission is as valid now as it was then, for in general every reader is already clear about what the term 'species' implies. It will be realized that the majority of the examples described in the preceding pages have been concerned with evolutionary changes occurring *below* the species level.

Modern advances in such areas as biochemistry, genetics, ecology, and palaeontology have thrown important new light on the criteria that can contribute towards decisions about the scope of species. These criteria can be summarized as follows:

(a) *Anatomical.* The members of a species can usually be expected to exhibit a clearly defined range of variation. Classificatory grouping is inevitably somewhat arbitrary, and where misfits occur one way of resolving the problem is to introduce sub-species. This is a practice particularly favoured by botanists, whereas zoologists tend to see species as more variable, 'polytypic' groups of organisms. Bearing in mind that much taxonomic work inevitably takes place in museums, anatomical criteria still play a large part in delineating species boundaries.

(b) *Geographical.* A species is normally limited to a distinct geographical range. For instance, the carrion crow, *Corvus corone,* is restricted in Europe to western and south-western areas, whereas the hooded crow, *C. cornix,* occurs mainly in the east and north.

(c) *Genetical.* Crosses between members of the same species are normally fertile and lead to the production of fertile offspring. Those between different species usually either fail or produce offspring which are themselves infertile, as generally occurs in the mule–the product of pairing a horse, *Equus caballus,* and a donkey, *E. asinus.* However, the ability of different species on occasions to cross under natural conditions and to produce fertile offspring is illustrated by our two species of crow mentioned in (b) above. Where their geographical ranges overlap, for instance in north Scotland, they hybridize freely, but in relation to their total range the degree of overlap between them is, in fact, quite small.

(d) *Ecological.* Members of the same species usually exhibit similar behaviour patterns and habitat preferences. Their basic ecological requirements are, therefore, much the same and it is rare to find two different species occupying the same ecological niche within a particular ecosystem.

(e) *Physiological.* Within a species there exists a limited and characteristic range of physiological and biochemical variation, such as the ability to produce a particular range of pigments or other metabolic by-products that may be of survival value. Since the biochemical pathways through which these substances are formed must often be of great antiquity, their occurrence frequently extends into larger taxonomic groups as well, such as the Family or Order, and therefore tend to be of limited value as criteria at the species level.

(f) *Palaeontological.* Species occupy a limited and distinct range in time. For most extinct forms, we have little or no knowledge of the ecological conditions in which they lived nor of the circumstances which eventually determined their downfall. In attempting to assign species status to fossils, it is therefore necessary to use tests rather different from those appropriate to living forms. One obvious parameter is size, but in applying it we are at once faced with the problem of not knowing how much of the observed variation is genetic and how much is environmental.

When assigning an organism to a particular species it will seldom be possible to apply more than two or three of the criteria outlined above. Although the modern procedure undoubtedly represents a step forward when compared with that of Darwin's time, it still leaves a great deal to be desired. The need for increased objectivity in taxonomic procedure has prompted an attempt to reduce classification entirely to quantitative terms (numerical taxonomy) by breaking down the features of organisms to be compared into a number of 'unit characters'. It is too early yet to say how successful this procedure will prove to be and whether the necessarily arbitrary nature of the characters chosen may not eventually lead to the same kinds of problems that arise at present from the more traditional approach. Further consideration is outside the scope of this book but references for further reading are included in the Bibliography.

In the long term, the most satisfactory way of judging the resemblances and differences between organisms may well be through a comparison of their respective genotypes. Recent discoveries in biochemical genetics (page 137) have already hinted at how this might be achieved–but only in the very distant future.

Evolution above species level

In discussing evolutionary changes, most of the examples quoted so far have taken place below the species level. Such changes are sometimes referred to collectively as micro-evolution, in contrast to those occurring at species level and above which are grouped collectively as macro-evolution. While such a division probably has little meaning in the analysis of the mechanisms involved, it nonetheless provides a useful means of categorizing the products, as we find them.

In spite of the incompleteness of the fossil record and the circumstantial nature of the evidence provided by other branches of biology, our knowledge of the broad outlines of animal and plant evolution is now reasonably complete. Such studies show that one of the chief features of the long history of living organisms has been their tendency to develop along certain quite distinct lines. In many instances these lines have led in the direction of obvious advantages which must have proved a benefit in the struggle for survival. On the other hand, some trends appear to have had no survival value whatever, and even in the end to have proved disadvantageous.

An example of the first is provided by the placental mammals of the late Eocene. Arising from small, unspecialized creatures, they branched out into a great variety of forms occupying a wide range of ecological niches. Thus from the primitive carnivores (Creodonta) there arose the seals (Pinnipedia) and the modern terrestrial forms, such as dogs, cats, and badgers (Fissipedia). Similarly, a stock resembling the early insectivores gave rise to the rodents and primates (Lemuroids and Tarsioids).

This process of *adaptive radiation,* in which a particular group spread out and its members become adapted to a variety of ecological conditions, can be followed along each successful evolutionary line, which eventually diverges further into a number of sub-lines. Thus the whales are regarded as having arisen from a carnivore stock closely related to the Creodonts. These subsequently branched out into three divergent lines: the whalebone whales, which are highly specialized for filter feeding; the great toothed whales, preying mainly on deep-sea molluscs; and the porpoises and dolphins, specialized for fish-eating.

There appear to have been many occasions on which evolutionary trends in related groups of organisms ran parallel to one another in a remarkable way. Thus, while the early placental mammals were evolving in other parts of the world, an odd group of primitive creatures was isolated in Australasia and South America. These were the marsupials, and they proceeded to radiate into a variety of types, some of them closely resembling those of the placentals. This is particularly true of the present-day marsupial mole (*Notoryctes*) and wolf (*Thylacinus*), which exhibit apparent *parallel evolution,* simulating their placental counterparts to a marked degree. Other forms have also come to resemble their corresponding placentals in a superficial way while occupying similar ecological niches, but the specializations involved have generally been achieved by different means. It is interesting to note that whenever human interference has resulted in competition between marsupials and placentals (as when the rabbit was introduced into Australia), the latter have always succeeded in replacing the former. This provides a measure of their relative degree of adaptability, and no doubt reveals the results of less severe selection under the isolated conditions in which the marsupials have lived.

It is not difficult to account for adaptive radiation on a Darwinian basis. Indeed, evolution of the kind just described is precisely what we would have expected if selection had worked for many millions of years on variable and tolerant species, faced with colonizing a diversity of habitats. Different regions, once invaded by a stock with a tendency to form divergent lines, were subsequently exploited by its offshoots, who adopted a variety of modes of life. The two phases–initial colonization and subsequent radiation–both essentially involved progressive adaptation.

A similar explanation can be used to account for apparent parallel evolution in related groups. Presumably these evolved from a common stock in the not very remote past. The ancestral mammals, for instance, are thought to have been small primitive forms ranging over Europe and North America in the late Cretaceous. These must all have possessed a genetic constitution with much the same *potentiality* for evolution. The course of their subsequent development was no doubt determined partly by individual variation and partly by the peculiar and diverse environments in which they found themselves.

Instances are sometimes cited in which lines of evolutionary development in distantly related groups appear to have converged, producing a similar result by independent means. Such *convergence* has been claimed, for example, in the evolution of the eyes of the vertebrates and Cephalopod molluscs, both of which possess a well-developed lens and retina (Figure 53). Now only four types of eye mechanism appear to be possible; the compound structure used by arthropods, and those corresponding to the pinhole camera, the lens camera, and the reflecting telescope. It is quite easy to picture a series of steps leading from one type of 'camera-eye' to the other; indeed, the series can easily be constructed from species living today. It is therefore not surprising that the same mechanism should have evolved several times. But the reflector would be of little use in its early stages, and has never appeared. Similarly, among animal respiratory pigments, the haemoglobins have been repeatedly developed quite independently by unrelated species when the necessity for increased oxygen transport has arisen. This is

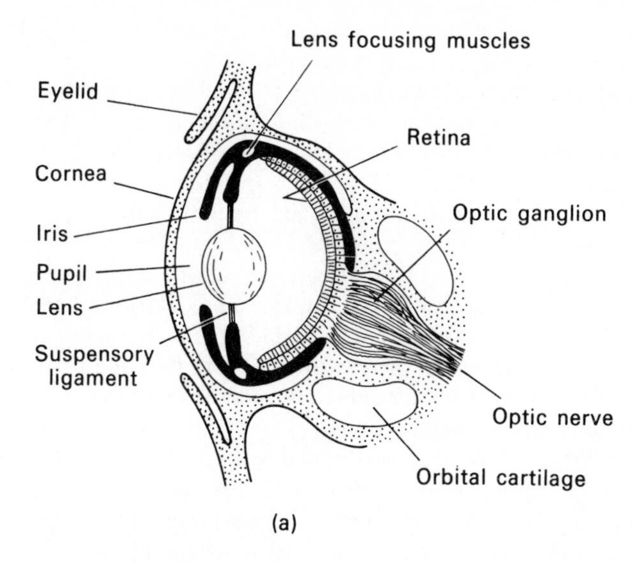

(a)

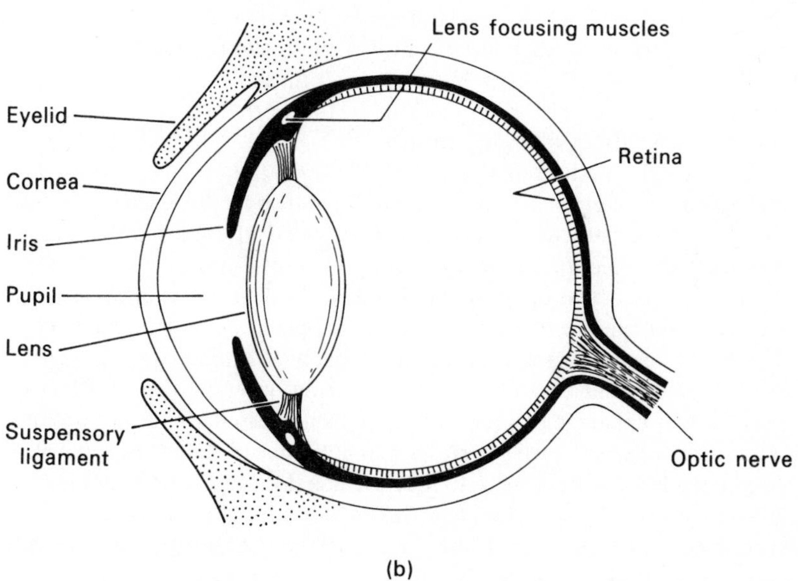

(b)

Figure 53. (a) Diagrammatic vertical section of the eye of a Cephalopod mollusc (squid) and (b) a similar view of the eye of a vertebrate (mammal)

understandable when we remember that all organisms possess the necessary precursor, an iron-porphyrin compound, in the intracellular respiratory system of cytochrome. Furthermore, such pigments need not be concerned with respiration at all, as in the nodules of leguminous plants where the role of haemoglobin is probably associated with nitrogen fixation.

Another characteristic evolutionary trend brought forward in the past as evidence against Darwinism, is *orthogenesis.* This is the tendency of many species to evolve in a particular direction, as if impelled by some overwhelming force to proceed towards a predetermined goal. The horse provides a well-known example, in which the reduction of the digits from four to one and an increase in size from that of a fox terrier to present-day forms took place progressively from the Eocene onwards (a period of about 40 million years).

A point which is often overlooked by exponents of orthogenesis is the fact that evolutionary series such as that of the horse never progress in a straight line, but invariably diverge into a number of side branches. The extent of this branching varies greatly in different groups. Thus, while the history of the horse is comparatively simple, that of the closely related rhinoceros is much more complex and includes a great number of divergent forms. Furthermore, a single group showing a relatively simple evolutionary history in one part of the world may radiate into a complex diversity of types in another.

At first sight, such examples suggest the operation of an evolutionary mechanism different from anything considered so far. But, viewed more closely, orthogenetic development is what we might expect in the long run if natural selection is operating as we believe. For once a variation leading in a particular direction has been established, the chances are greatly increased of a further advance taking place along the same lines. This is not because the new type is more likely to mutate again in the same direction, but because a similar mutation, when it occurs, will have the effect of enhancing a process already begun. Similarly, mutations in other directions will have a correspondingly reduced chance of survival. As Huxley has put it, 'a specialized line finds itself at the bottom of a groove cut for it by selection; and the further a trend towards specialization has proceeded, the deeper will be the biological groove in which it has thus entrenched itself'.

More difficult to understand are the numerous occasions on which orthogenetic evolution appears to have carried a group of organisms past the point at which they could have derived any benefit from their new characteristics. In other words, evolution seems eventually to have become non-adaptive! Thus, there is no doubt that the development of giant bodily size and huge horns were the prelude to extinction in numerous separate groups of the great hoofed mammals (Titanotheres) which existed during the Oligocene. Similarly, the Irish elk, *Cervus giganteus,* possessed antlers nearly 6 m across, which may well have proved to be an encumbrance rather than an asset. One group of molluscs, the ammonites (relatives of the present-day cuttlefish and octopus), possessed shells divided internally by plates (septa) into a number of separate chambers. Where the edges of the septa appear on the surface of the shell they are known as sutures. At the time of their first appearance, in the Ordovician, the shells were long and straight, the sutures between the chambers being relatively simple and uncurved (Figure 54). Subsequent evolution resulted in the coiling of the shells and a great increase in the complexity of the suture lines, which became wavy in outline (Figure 54). Since we know virtually nothing of the functional significance of these changes, nor of the ecological conditions which prevailed at the time, it is impossible to say whether such variations in structure could have been adaptive or not.

One of the remarkable features of ammonite history is that, having embarked on modifications in such specific directions, a number of groups then proceeded to reverse the process, the shells becoming uncoiled and the suture lines simplified once more. This trend towards the primitive condition invariably proved to be the prelude to extinction. It is particularly noticeable in fossils from the late Trias—the period when the majority of the great ammonite groups died out.

Extreme examples such as this are rare, and in the absence of any knowledge of the environmental conditions it is difficult to judge whether they provide an exception to Darwinism or not. We have seen (page 70) how an advantageous characteristic in an animal or plant must often involve a balance between good and bad. It is thus possible that the radical changes which took place in ammonite structure may have been associated with other beneficial characteristics, possibly of a physiological kind, of which we know nothing. At the present, however, this must remain mere speculation.

Can non-Darwinian evolution occur?

Some of the most fundamental changes in the outlook of biologists during recent years have stemmed from advances in biochemistry, which

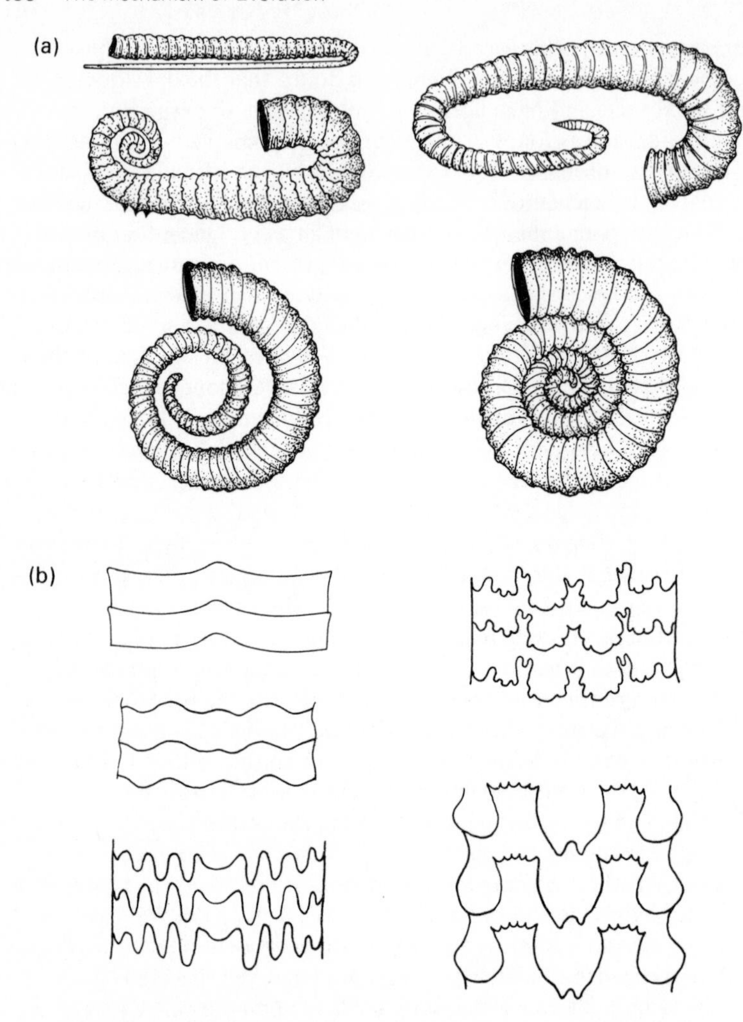

Figure 54. Stages in the evolution of ammonite shells (diagrammatic): (a) degrees of coiling and (b) variation in complexity of the suture lines. Subsequent uncoiling and simplification of the suture lines proved to be the prelude to extinction

have influenced thinking throughout the whole subject. One outcome has been that subjects which previously tended to be somewhat isolated from each other are now rapidly achieving a basic unity. This applies particularly to such areas as biochemistry, physiology, ecology, genetics, and evolution. As has become apparent many times in the preceding pages, the ultimate explanation of adaptation is, more often than not, to be found in biochemical terms, so that any movement of thinking and research in this direction is to be welcomed.

A particular instance of such a change in outlook is to be found in our approach to mutation. In Chapter 2 (page 35) it was stated that gene mutation, in its classical sense, is rare, the rate being on average of the order of 1 in 1 000 000 individuals. Micro-organisms such as bacteria and viruses have proved to be valuable material for the study of mutation since they multiply rapidly, enabling huge numbers to be examined in a short period of time. New variants which are rare in normal breeding experiments can therefore be detected quite frequently. Moreover, the study of such mutants, many of whose effects are biochemical, has led to extensive mapping of the DNA chain and the identification of the mutational units (mutons). From such work it now seems reasonably certain that each muton represents a single nucleotide. Adjacent mutons may interact with one another and produce the same or different effects. Those that act together represent functional units of genetic material and are known as cistrons.

Investigation of the rates of mutation of particular mutons in micro-organisms have shown these to be far higher than classical beliefs would suggest—possibly by a new order of magnitude. This is also true of some other organisms, as Stormont has shown in cattle for the locus controlling the blood group β-system, where the incidence of mutation-like events is of the order of 1 in 500. In the brook trout, Wright and Atherton have identified a mutation concerned with the enzyme lactate dehydrogenase with a frequency of 1 in 50. Evidently, at molecular level there is a large and continuous accretion of biochemical variants resulting from high rates of mutation and leading to a wide range of biochemical polymorphisms, many of which undoubtedly remain undetected. How much of this genetic variation is subject to natural selection is hard to say at present, but it could be, as Dyer has suggested, that by virtue of their molecular nature and their minimal degree of influence on the organism as a whole, many of the new mutations are selectively neutral, their incorporation in the biochemical make-up of the plants and animals concerned being therefore of a non-Darwinian kind. But until we know more about the fundamental effects of such variations on the organisms that carry them, particularly in relation to their possible survival value, the question as to whether or not their existence can be accounted for in selectionist terms must inevitably remain unanswered.

Conclusion

From the foregoing account it will, I hope, be clear that, while there are various types of evolutionary change which still remain unexplained, many other forms of adaptation can be accounted for in terms of survival value. This is the essence of the modern theory of evolution. Moreover, we now know of a workable evolutionary mechanism which can be tested by experiment.

Advances in genetics dating from the time of Mendel have provided a means of explaining variation in animals and plants. Darwin believed in blending inheritance—that is to say, the tendency of genetic factors derived from opposite parents to merge with one another in the cells of the offspring. This process, as he clearly realized, could only result in a progressive reduction of variability and the loss of beneficial genes by dilution with others. Consequently, he was driven to postulate a high rate of mutation in order to offset the ill effects of blending.

With a knowledge of Mendelian inheritance came the realization that genes are permanent entities, except in so far as they are subject to mutation. This, coupled with the fact of their random assortment in the nuclei of the germ cells (subject to limitations imposed by linkage, crossing-over and affinity), at once overcame the difficulties inherent in the blending theory. At first, study of the influence of specific genes was confined mainly to structural characteristics, such as the colour of flowers or the shape of the comb in poultry. One of the features of modern research has been an inquiry into inherited physiological and biochemical effects, many of which have been mentioned in the previous chapters.

An aspect of gene-action which is particularly significant in evolution is its influence on growth rate. We still have a great deal to learn about the precise way in which this complex process is controlled, but there is no doubt that the potentiality for growth in many organisms (possibly in all) is inherited and subject to the influence of natural selection. This may determine not only the extent of development of a particular organ, but also the relative time at which it appears in the life cycle.

Some notable examples of the influence of growth rate in evolution are to be found among the vertebrates, in many of which the cessation of growth is determined by the attainment of sexual maturity. A group of newts (commonly known as axolotls) have achieved sexual development while retaining the bodily characteristics of normal newt larvae, such as external gills and disproportionately large heads. One of these, *Siredon axolotl,* if fed on thyroxin (the hormone secreted by the thyroid gland), can be induced to metamorphose, giving rise to a normal newt (*Amblystoma tigrinum*). This phenomenon, involving the precocious development of the reproductive organs and a relative reduction in the speed of somatic growth, is known as *neoteny.*

There is some evidence that this process may have played a significant part in the evolution of man. For instance, many human features, such as the relatively large brain and the absence of brow ridges (Figure 55), are

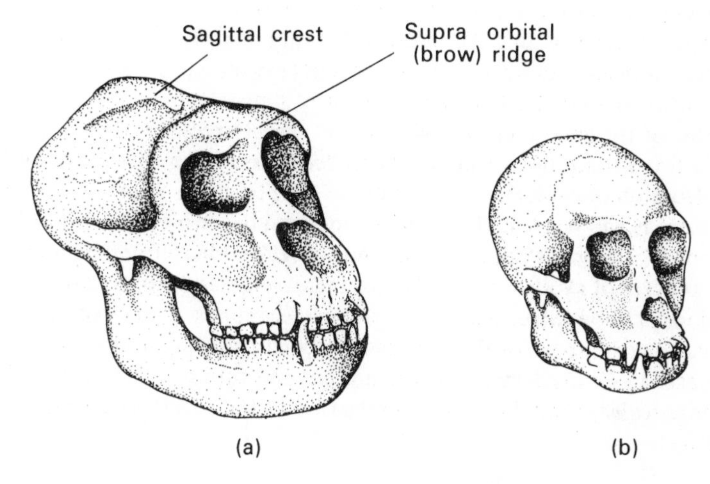

Figure 55. Skulls of (a) old and (b) young gorilla. Note in the young skull the lack of a sagittal crest, the more vertical face and the relative absence of brow ridges, features which are characteristic of the human skull. (After Reynolds.)

characteristic of the young stages of apes such as the gorilla. The resemblance extends further to the growth rate of the brain relative to that of the rest of the body, and also to the enormous size of the cerebral hemispheres at birth compared with that of neighbouring regions.

Another result of the relative slowing-down of bodily development may be that the appearance of certain structures is so pressed back in time that they are reduced to vestiges. Such *retardation* accounts for the evolution of our 'wisdom teeth' (vestigial molars), which are often never cut at all. The lack of hair over the general body surface in human beings is another characteristic probably attributable to the same process.

Other notable advances in evolutionary studies during recent years have resulted from improvements in experimental techniques, and in particular from a more precise type of mathematical treatment. Thus the various methods of marking individuals have enabled the spread of genes with detectable phenotypic effects to be studied in the wild state. Polymorphic species have proved invaluable for this kind of work. At the same time, the increased accuracy with which we can now determine the density of living organisms has enabled more exact estimates to be made of the magnitude of selective advantages and disadvantages. Among polymorphic species such as the peppered moth and in those with polygenic variation like the meadow brown butterfly, selection pressures operating against disadvantageous phenotypes have proved to be of a magnitude far greater than was previously supposed.

One of the great merits of the present 'neo-Darwinian' theory of evolution is that, unlike any of its predecessors, it provides a workable explanation of evolution susceptible of scientific test. From this brief account it will be clear that experimental studies, which are still in their infancy, offer great opportunities, not only for a better understanding of the process of evolution but also for increasing our fund of basic biological knowledge. With the universal acceptance of the fact of evolution, the focus of thought has switched from a concentration on the past to a consideration of the present and future. The result has been a more realistic and dynamic appreciation of the nature of evolutionary change.

Glossary

Adaptive Radiation. The evolution of a group of organisms along a number of different lines involving adaptation to a variety of ecological conditions.

Affinity. The tendency of certain bivalent chromosomes not to orientate themselves at random in relation to the poles of the cell at the first meiotic prophase. The result is non-random segregation and a further exception to the Law of Independent Assortment.

Alleles (Allelomorphs). Alternative forms of a gene which occupy identical positions on homologous chromosomes.

Aposematic Colours. Those which advertise the presence of an organism to potential predators to denote the inadvisability of attack.

Autosome. A chromosome other than a sex chromosome.

Barrier. Any ecological factor which restricts the increase in range of a species.

Biotic Factors. Biological factors resulting from the interaction of living organisms with one another, such as food and population density.

Chimaera. An organism whose tissues are of two or more kinds which are genetically different. It can occur either as a result of somatic mutation or from grafting together two different plants whose characteristics become mixed (*graft hybrid*).

Chromatids. Half chromosomes, i.e. the bodies produced in pairs by the longitudinal splitting of chromosomes prior to nuclear division. Each chromatid becomes a chromosome in the resulting daughter nucleus.

Chromatin. The material which the chromosomes contain.

Chromosome. One of the deeply staining paired structures present in a nucleus. Their number is constant in each species and they are the carriers of genes.

Cistron. A group of adjacent nucleotides, mutation in any one of which affects the same character. Functional unit of genetic material.

Climatic Factors. Biological factors which are mainly physical in nature, such as temperature and light.

Cline. A graded series of changes in an organism, either structural or physiological, taking place gradually within a particular zone.

Codon. A sequence of three bases along a DNA chain which represent the code of a single amino acid.

Convergence. The tendency of distantly related groups of organisms to evolve similar structures by independent means.

Crossing-over. An interchange of groups of genes between the members of a homologous pair of chromosomes.

Cryptic Colours. Those which tend to match the environment in which an animal lives.

Cultivar. A domesticated strain of plants.

Cytochrome. A collective term used to describe a mixture of enzymes concerned in respiration and responsible for controlling oxygen consumption in aerobic organisms.

Cytoplasm. The living substance of the cell excluding the nucleus.

Dimorphism. The occurrence together of two forms of the same species.

Diploid Cells. Those possessing both members of each chromosome pair (often written as $2n$).

DNA (*Desoxyribonucleic acid*). The nucleic acid found mainly in chromosomes which constitutes the substance of genes.

Dominant. A character as fully developed when the alleles determining it are heterozygous as when they are homozygous (see *Recessive*).

Ecology. The relationship of organisms to one another and their environment.

Environment. The sum total of the conditions in which an organism lives.

Environmental Variation. Variation in a character of an organism due to external influences and occurring irrespective of any alteration in its hereditary constitution (see *Genetic Variation*).

Epigamic Colours. Those occurring in animals for the purpose of courtship and display to the opposite sex.

Evolution. Changes in the characteristics of organisms occurring in successive generations in response to the environment in which they live.

Fossil. Remains of an organism preserved in some natural medium such as rock, amber, or peat.

Geiger-Müller Counter. An instrument for the detection of ionizing radiations, chiefly α, β, and γ rays.

Gene Complex. The sum total of the genetic factors of an organism interacting to produce an internal environment in which all the genes must operate.

Gene-frequency. The numbers of a particular gene present in a population expressed as a percentage of the total number of loci available for them to occupy.

Genes. The hereditary units. They are responsible for the production of a given set of characters in any particular environment.

Genetics. The study of heredity and variation.

Genetic Variation. Variation in a character of an organism recurring as a result of mutation or recombination of genes.

Genotype. The gene-complement carried by an organism (see *Phenotype*).

Half-life. The time taken for the activity of a radioactive element to decay to one-half of its original value.

Haploid Cells. Those possessing only one member of each chromosome pair, e.g. gametes (often written as n).

Heritability. The extent to which a particular character is determined by inheritance. The ratio of the genotypic variance to the observed phenotypic variance.

Heterozygote. An individual in which the members of a gene pair (alleles) are dissimilar.

Homozygote. An individual in which the members of a gene pair (alleles) are similar.

Hormones. Organic compounds of great physiological importance which are essential in small quantities for the normal functioning of organisms. They are secreted by special glands (endocrine) or tissues.

Isotopes. Atoms of the same element but differing in atomic weight.

Linkage. The tendency for certain genes to remain together instead of assorting independently, since they are carried on the same chromosome. Linkage thus provides an exception to Mendel's Second Law.

Melanic Form. An animal whose colour is brown or black on account of the pigment melanin.

Melanin. A nitrogenous animal pigment with a black or brownish colour.

Multiple Alleles. A series of genes occurring at the same locus on a chromosome which have arisen by mutation.

Mutation. The inception of a heritable variation.

Neoteny. Precocious sexual development accompanied by a relative reduction in the rate of bodily growth. The process is almost certainly under genetic control and has played an important part in evolution.

Nucleus. That portion of a cell which contains the chromosomes. It is essential for the life of all normal cells and is perpetuated by division.

Orthogenesis. The tendency for related groups of organisms to evolve in the same direction. This can be explained on a Darwinian basis and there is no evidence of a 'directional urge'.

Phenocopy. A structure whose development is inherited but which can also be simulated by environmental effects alone, e.g. the thickness of the epidermis on the heel of the foot.

Phenotype. The characteristics of an organism resulting from the reaction of a given genotype with a particular environment (see *Genotype*).

Physiology. The study of the various vital processes occurring in living organisms.

Plastids. Bodies carried in the cytoplasm of plant cells. They are capable of multiplication but are distributed irregularly at each cell division.

Polymorphism. The occurrence together of two or more forms of the same species, the rarest existing with a frequency above that of recurrent mutation.

Polyploidy. Circumstances in which more than the two members of each gene-pair are present in the cells of an organism.

Pure-line. The descendants from an individual in which self-fertilization has occurred. All the genes are, therefore, in the homozygous state.

Quadrat. A square of known size, used to determine the distribution and density of plants in different localities.

Recessive. A character produced only when the alleles controlling it are homozygous (see *Dominant*).

Retardation. Reduction in the rate of growth of a structure relative to that of others. The process may result in the formation of vestiges, e.g. human 'wisdom teeth' (vestigial molars).

Sex-chromosomes. The X and Y chromosomes.

Sex-controlled Inheritance. Characteristics which are only apparent in a particular sex. The genes concerned may be carried in the sex chromosomes or in the autosomes.

Sex-linked genes. Genes carried in the sex-chromosomes.

Somatic Mutation. Mutation taking place in the somatic (body) cells of an organism instead of in those which form gametes.

Specialization. Structural or physiological adaptation by an organism to a particular set of environmental conditions.

Spot-average. A parameter for estimating second-order variation in adults of the butterfly *Maniola jurtina,* obtained by multiplying the class values by their frequencies, summating, and dividing by the total.

X chromosome. The chromosome carrying the genes which control the determination of sex.

Y chromosome. The partner of the X chromosome in one of the two sexes (generally the male). It contains few genes and plays little part in sex-determination.

Zygote. The first cell of a new organism resulting from the fusion of two gametes.

Bibliography

Books for further reading and reference

Books marked (A) *are advanced accounts. Those marked* (G) *are suitable for general reading.*

(G) Berry, R. J. (1965). *Genetics,* English Universities Press.
(G) Briggs, D. and Walters, S. M. (1969). *Plant Variation and Evolution,* Weidenfeld and Nicolson.
(G) Campbell, B. G. (1967). *Human Evolution,* Heinemann.
(G) Clarke, C. A. (1970). *Human Genetics and Medicine,* Arnold.
(A) Cook, L. M. (1971). *Coefficients of Natural Selection,* Hutchinson.
(A) Creed, E. R. (editor) (1971). *Ecological Genetics and Evolution,* Blackwell.
(G) Crowson, R. A. (1970). *Classification and Biology,* Heinemann.
(G) Darwin, C. (1872, 6th edn. reprinted 1956). *The Origin of Species,* Oxford.
(G) De Beer, G. R. (1963). *Charles Darwin,* Nelson.
(A) Dobzhansky, T. (1951, 3rd edn.). *Genetics and the Origin of Species,* Columbia.
(G) Dobzhansky, T. (1965). *Heredity and the Nature of Man,* Allen and Unwin.
(G) Dowdeswell, W. H. (1966, 2nd edn.). *Introduction to Animal Ecology,* Methuen.
(A) Fisher, R. A. (1930). *The Genetical Theory of Natural Selection,* Oxford.
(G) Ford, E. B. (1965). *Genetic Polymorphism,* Faber and Faber.
(G) Ford, E. B. (1967, 7th edn.). *Mendelism and Evolution,* Methuen.
(G) Ford, E. B. (1968, 6th edn.). *Genetics for Medical Students,* Methuen.
(A) Ford, E. B. (1971, 3rd edn.). *Ecological Genetics,* Chapman and Hall.
(A) Lewis, K. R. and John, B. (1964). *The Matter of Mendelian Heredity,* Churchill.
(G) Mather, K. (1966). *Human Diversity,* Oliver and Boyd.
(G) Maynard Smith, J. (1969, 2nd edn.). *The Theory of Evolution,* Penguin.
(G) Mayr, E. (1964). *Systematics and the Origin of Species,* Dover.
(G) Medawar, P. (1960). *The Future of Man,* Methuen.
(G) Olby, R. C. (1966). *Origins of Mendelism,* Constable.
(G) Sheppard, P. M. (1967, 2nd edn.). *Natural Selection and Heredity,* Hutchinson.
(G) Simpson, G. G. (1964). *This View of Life,* Harcourt, Brace and World.
(G) Sullivan, N. (1968). *The Message of the Genes,* Routledge and Kegan Paul.
(G) Vorzimmer, P. J. (1972). *Charles Darwin: The Years of Controversy,* University of London Press.

Selected references to original papers

Aird, I., Bentall, H. H., Fraser Roberts, J. A. (1953). 'A relationship between cancer of the stomach and the ABO blood groups', *British Medical Journal,* **2,** 799-801.

Allen, W. R. and Sheppard, P. M. (1971). 'Copper tolerance in some Californian populations of the monkey flower, *Mimulus guttatus*', *Proceedings of the Royal Society, B,* **177**, 177-96.

Allison, A. C. (1954). 'Protection afforded by the sickle-cell trait against subtertian malarial infection', *British Medical Journal,* **1**, 290.

Beaufoy, E. M. and S., Dowdeswell, W. H. and McWhirter, K. G. (1970). 'Evolutionary studies on *Maniola jurtina*: the Southern English stabilization, 1961-8', *Heredity,* **25**, 105-12.

Bishop, J. A. and Harper, P. (1970). 'Melanism in the moth, *Gonodontis bidentata*: a cline within the Merseyside conurbation', *Heredity,* **25**, 449-56.

Bodmer, W. F. (1958). 'Natural crossing between homostyle plants of *Primula vulgaris*', *Heredity,* **12**, 363-70.

Bodmer, W. F. (1960). 'The genetics of homostyly in populations of *Primula vulgaris*', *Philosophical Transactions of the Royal Society, B,* **242**, 517-49.

Bradshaw, A. D., McNeilly, T. S. and Gregory, R. P. G. (1965). 'Industrialisation, evolution and the development of heavy metal tolerance in plants', *5th Symposium of the British Ecological Society,* Blackwell.

Brower, L. P. (1969). 'Ecological chemistry', *Scientific American,* **220**, 22-9.

Cain, A. J. and Currey, J. D. (1968). 'Climate and selection of banding morphs in *Cepaea* from the climate optimum to the present day', *Philosophical Transactions of the Royal Society, B,* **253**, 483-97.

Cain, A. J. and Currey, J. D. (1968). 'Ecogenetics of a population of *Cepaea nemoralis* subject to strong area effects', *Philosphical Transactions of the Royal Society, B,* **253**, 447-82.

Cain, A. J. and Sheppard, P. M. (1950). 'Selection in the polymorphic land snail *Cepaea nemoralis*', *Heredity,* **4**, 274-94.

Cameron, T. W. M. (1952). 'Parasitism, evolution and phylogeny', *Endeavour,* **11**, 193-99.

Camin, J. H. and Ehrlich, P. R. (1959). 'Natural selection in water snakes (*Natrix sipedon*) on islands in Lake Erie', *Evolution,* **2**, 504-11.

Clarke, B. (1960). 'Divergent effects of natural selection in two closely related polymorphic snails', *Heredity,* **14**, 423-43.

Clarke, C. A. *et al.* (1955). 'The relation of the ABO blood groups to duodenal and gastric ulceration', *British Medical Journal,* **2**, 643-6.

Clarke, C. A. and Sheppard, P. M. (1966). 'A local survey of the distribution of industrial melanic forms in the moth *Biston betularia,* and estimates of the selective value of these in an industrial environment', *Proceedings of the Royal Society, B,* **165**, 424-39.

Cook, L. M., Askew, R. R. and Bishop, J. A. (1970). 'Increasing frequency of the typical form of the peppered moth in Manchester', *Nature,* **227**, 1155.

Cook, L. M., Brower, L. P. and Croze, H. J. (1969). 'The accuracy of a population estimation from multiple recapture data', *Journal of Animal Ecology,* **36**, 57-60.

Crawford-Sidebotham, T. J. (1972). 'The role of slugs and snails in the maintenance of the cyanogenesis polymorphisms of *Lotus corniculatus* and *Trifolium repens*', *Heredity,* **28**, 405-12.

Creed, E. R., Dowdeswell, W. H., Ford, E. B. and McWhirter, K. G. (1970). 'Evolutionary studies on *Maniola jurtina.* The "boundary phenomenon" in Southern England 1961-68', *Essays in Evolution and Genetics,* 263-87 (ed. M. K. Hecht and W. C. Steere), Appleton-Century-Crofts.

Creed, E. R., Ford, E. B. and McWhirter, K. G. (1964). 'Evolutionary studies on *Maniola jurtina*: the Isles of Scilly 1958-59', *Heredity,* **19**, 471-88.

Crosby, J. L. (1959). 'Outcrossing in homostyle primroses', *Heredity,* **13**, 127-31.

Daday, H. (1954). 'Gene frequencies in wild populations of *Trifolium repens.* I. Distribution by latitude, *Heredity,* **8**, 61-78.

Daday, H. (1965). 'Gene frequencies in wild populations of *Trifolium repens.* IV. Mechanism of natural selection', *Heredity,* **20**, 355-65.

Day J. C. L. and Dowdeswell, W. H. (1968). 'Natural selection in *Cepaea* on Portland Bill', *Heredity,* **23**, 169-88.

Dobzhansky, T. and Pavlovsky, O. (1957). 'An experimental study of interaction between genetic drift and natural selection', *Evolution,* **11**, 311-19.

Dowdeswell, W. H. (1956). 'Isolation and adaptation in populations of the Lepidoptera', *Proceedings of the Royal Society,* **145**, 322-9.

Dowdeswell, W. H. (1961). 'Experimental studies on natural selection in the butterfly *Maniola jurtina*', *Heredity,* **16**, 39-52.

Dowdeswell, W. H. and McWhirter, K. G. (1967). 'Stability of spot-distribution in *Maniola jurtina* throughout its range', *Heredity,* **22**, 187-210.

Dyer, K. F. (1971). 'The quiet revolution. A new synthesis of biological knowledge', *Journal of Biological Education,* **5**, 15-24.

Fisher, J. (1952). *The Fulmar,* Collins, New Naturalist Series.

Fisher, R. A. and Ford, E. B. (1950). 'The "Sewall Wright effect" ', *Heredity,* **4**, 117-19.

Ford, E. B. (1958). 'Darwinism and the study of evolution in natural populations', *Proceedings of the Linnaean Society. Zoology,* **44**, 41-8.

Ford, E. B. (1960). 'Evolution in progress', From *Evolution after Darwin* (ed. Sol. Tax.), **1**, 181-96, Chicago University Press.

Ford, E. B. (1961). 'The theory of genetic polymorphism, *Symposia of the Royal Entomological Society of London,* **1**, 11-19.

Ford, E. B. and Sheppard, P. M. (1969). 'The *medionigra* polymorphism of *Panaxia dominula*', *Heredity,* **24**, 561-9.

Ford, H. D. and E. B. (1930). 'Fluctuation in numbers and its influence on variation in *Melitaea aurinia*', *Transactions of the Entomological Society of London,* **78**, 345-51.

Foulds, W. and Grime, J. P. (1971). 'The influence of soil moisture on the frequency of cyanogenic plants in populations of *Trifolium repens* and *Lotus corniculatus*', *Heredity,* **28**, 143-6.

Gregory, R. P. G. and Bradshaw, A. D. (1965). 'Heavy metal tolerance in populations of *Agrostis tenuis* and other grasses', *New Phytologist,* **64**, 131-43.

Haldane, J. B. S. (1952). 'Variation', *New Biology,* **12**, 9-26.

Haskell, G. (1951). 'Plant chromosome-races and their ecology in Great Britain', *Nature,* **167**, 628-9.

Hovanitz, W. (1944). 'The distribution of gene frequencies in wild populations of *Colias', Genetics,* **29**, 31-60.

Jones, D. A. (1972). 'Cyanogenic glucosides and their function.' In *Phytochemical Ecology* (editor J. B. Harborne), 7, 103-22, Academic Press.

Jones, D. A. (1970). 'On the polymorphism of *Lotus corniculatus.* III Some aspects of selection', *Heredity,* **25**, 633-41.

Karn, M. N. and Penrose, L. S. (1952). 'Birth weight and gestation time in relation to maternal age, parity and infant survival', *Annals of Eugenics,* **16**, 147-64.

Kettlewell, H. B. D. (1952). 'Use of radioactive tracers in the study of insect populations (Lepidoptera)', *Nature,* **170**, 584.

Kettlewell, H. B. D. (1956). 'Further selection experiments on industrial melanism in the Lepidoptera' , *Heredity,* **10**, 287-301.

Kettlewell, H. B. D. (1958). 'A survey of the frequencies of *Biston betularia* and its melanic forms in Great Britain', *Heredity,* **12**, 51-72.

Kettlewell, H. B. D. (1961). 'Selection experiments on melanism in *Amathes glareosa* (Esp.)', *Heredity,* **16**, 415-34.

Kettlewell, H. B. D. and Berry, R. J. (1969). 'Gene flow in a cline', *Heredity,* **24**, 1-14.

Kimura, M. (1968). 'Evolutionary rate at the molecular level', *Nature,* **217**, 624-6.

King, J. L. and Jukes, T. H. (1969), 'Non-Darwinian evolution', *Science,* **164**, 788-98.

Lack, D. (1965). 'Evolutionary ecology', *Journal of Animal Ecology,* **34**, 223-31.

Lamotte, M. (1959). 'Polymorphism of natural populations of *Cepaea nemoralis*', *Cold Spring Harbor Symposia on Quantitative Biology,* **24**, 65-85.

McWhirter, K. G. (1969). 'Heritability of spot-number in Scillonian strains of the meadow brown butterfly (*Maniola jurtina*)', *Heredity,* **24**, 314-18.

McWhirter, K. G. and Scali, V. (1966). 'Ecological bacteriology of the meadow brown butterfly', *Heredity,* **21**, 517-21.

Medawar, P. B. (1951). 'Problems of adaptation', *New Biology,* **11**, 10-26.

Millicent, E. and Thoday, J. M. (1961). 'Effects of disruptive selection. IV Gene-flow and divergence', *Heredity,* **13**, 205-18.

Murray, J. (1963). 'The inheritance of some characters in *Cepaea hortensis* and *Cepaea nemoralis* (Gastropoda)', *Genetics,* **48**, 605-15.

Owen, D. F. (1965). 'Density effects in polymorphic land snails', *Heredity,* **20**, 312-15.

Parsons, J. and Rothschild, M. (1964). 'Rhodanese in the larva and pupa of the common blue butterfly, *Polyommatus icarus*', *Entomologist's Gazette,* **15**, 58-9.

Richmond, R. C. (1970). 'Non-Darwinian evolution. A critique', *Nature,* **225**, 1025-8.

Sheppard, P. M. (1951). 'Fluctuations in the selective value of certain phenotypes in the polymorphic land snail *Cepaea nemoralis*', *Heredity,* **5**, 125-34.

Sheppard, P. M. (1952). 'Natural selection in two colonies of the polymorphic land snail *Cepaea nemoralis*,' *Heredity,* **6**, 233-8.

Sheppard, P. M. (1959). 'Blood groups and natural selection', *British Medical Bulletin,* **15**, 134-9.

Sheppard, P. M. (1969). 'Evolutionary genetics of animal populations: The study of natural populations', *Proceedings of the XIIth International Congress of Genetics,* **3**, 261-79.

Sumner, F. B. (1930). 'Genetic and distributional studies of three sub-species of *Peromyscus*', *Journal of Genetics,* **23**, 276-376.

Thoday, J. M. and Boam, T. B. (1959). 'Effects of disruptive selection. II Polymorphism and divergence without isolation', *Heredity,* **13**, 205-18.

Turner, J. R. G. and Williamson, M. H. (1968). 'Population size, natural selection and genetic load', *Nature,* **218**, 700.

Waddington, C. H. (1953). 'Epigenetics and evolution', *S.E.B. Symposia, 7 (Evolution),* 186-99.

Wolda, H. (1969). 'Stability of a steep cline in morph frequencies of the snail *Cepaea nemoralis*', *Journal of Animal Ecology,* **38**, 623-33.

Woolf, B. (1954). 'On estimating the relation between blood groups and disease', *Annals of Human Genetics,* **19**, 251-3.

Wright, Sewall (1948). 'On the roles of directed and random changes in gene frequency in the genetics of populations', *Evolution,* **2**, 279-94.

Index